Religion and the Technological Future

"Like weather reporters checking our daily atmospheric pressure, for nearly two decades Mercer and Trothen have been monitoring biohacking at the frontier of the human and the posthuman. They forecast a coming storm of theological and ethical conundrums. *Religion and the Technological Future* tells us how to ready ourselves for the storm."

—Ted Peters
Distinguished Research Professor of Systematic Theology and Ethics
Center for Theology and the Natural Sciences, Graduate Theological Union, USA
Co-editor, *Theology and Science*

"It is vital that we begin to analyze these difficult questions before, not after, they become reality."

—Bill McKibben
Schumann Distinguished Scholar
Middlebury College, USA
Author of *Falter: Has the Human Game Begun to Play Itself Out?*

"Human yearning to prevent morbidity and mortality is so intense that we mostly put it out of our minds, in any way that works—including religion, for some people. But this is just one of many ways to relate to our future and our higher powers' control over it. This outstanding textbook comprehensively surveys the ecosystem of such thinking. Whatever your stance on these issues, you'll find extensive food for thought here."

—Aubrey de Grey
Chief Science Officer, SENS Research Foundation

"Mercer and Trothen have laid down the intellectual infrastructure to enable theology to make a smooth transition into the transhuman future. And precisely because the book is firmly grounded in the religious world-view, it should also prove attractive to secular ethicists who approach their topic from a deep metaphysical perspective that wonders how the ongoing developments in science and technology are transforming what it means to be 'human.'"

—Steve Fuller
Auguste Comte Chair in Social Epistemology
University of Warwick, UK
Author of *Humanity 2.0*

"*Religion and the Technological Future* is the first comprehensive and deeply-considered textbook on the intersection of religious thought and twenty-first century technology. Its lively prose brings to life the intellectual, political, and ethical debates that frame our progress into the next stage of human culture. Mercer and Trothen take their years of leadership in the study of radical enhancement and transform that into a textbook that will inspire thoughtful engagement from students, scholars, and the general public."

—Robert M. Geraci
Professor of Religious Studies
Manhattan College, USA
Author of *Apocalyptic AI: Visions of Heaven in Robotics, Artificial Intelligence, and Virtual Reality*

"This timely textbook provides a valuable overview of the profound challenges and potential benefits that contemporary technology poses to humanity. Written by pioneering scholars in the field of religion and technoscience, this book introduces key concepts, themes, contributors, and debates on human enhancement and biohacking, radical life extension and cryonics, Artificial Intelligence and digital immortality. Intended for college-level courses in ethics, science and technology studies, religious studies, and psychology, the book is also most useful for programs in faith communities that ponder the promise and peril of technology."

– Hava Tirosh-Samuelson
Regents Professor of History
Director, Center for Jewish Studies
Irving and Miriam Lowe Professor of Modern Judaism
Arizona State University, USA

"*Religion and the Technological Future: An Introduction to Biohacking, Artificial Intelligence, and Transhumanism* invites thinkers from all disciplines and faiths to contemplate our collective future, and asks us: what sort of future do we want to make? Covering emerging and futuristic technologies, religious and ethical responses, and identifying areas of further inquiry, this ambitious volume offers clear, expert guidance through this urgently needed conversation about what we value about our human selves, and desire for our (post)human future."

—J. Jeanine Thweatt
School of Humanities and Sciences
Flagler College, USA

"The world is changing. A.I., robots, genetics, cybernetics, and many other areas demonstrate that traditional religious questions of 'what does it mean to be human?' and 'what is the role of religion?' are all being, or will be, rethought. Religious Studies should not be on the sidelines for this. Trothen and Mercer have provided a Religious Studies textbook for the 21st century that takes on these cutting edge issues and brilliantly shows the way religion is a full participant in these advances. As Religious Studies seeks to be relevant in this new world, this textbook may well become the new standard for introductory classes in Religion and Theology."

—Randall Reed
Professor of Religion
Appalachian State University, USA

"Faith, science, life, and technological power are forever intertwined. Since human-driven technological change is speeding up and the potency of life technologies are multiplying exponentially, pastors, religious leaders, and people of faith will need guides to recognize the perils and examine the promises. This guide will be indispensable for speaking love and truth and justice to the human condition of today and tomorrow."

—Pastor Peter Noteboom
General Secretary, The Canadian Council of Churches

"There are many books assessing some aspect of transhumanism, and more broadly the technological transformation of human beings, from a religious perspective. But to date no comprehensive overview has been on offer. That void has now been filled by an informed, accessible, and highly engaging text that deserves to be read and studied by a wide audience."

—Brent Waters
Stead Professor of Christian Social Ethics
Director, Stead Center for Ethics and Values
Garrett-Evangelical Theological Seminary, USA

"The future is coming sooner than we think. Our descendants may be a new species—*homo technicus*—as different from homo sapiens as we are from our ancestors. Trothen and Mercer provide a much needed exploration of what this new species might be like and how the crucial aspects of life—to be born, live, love, and die—will change. It is we who will determine the direction of those changes."

—Noreen Herzfeld
Reuter Professor of Science and Religion
St. John's University and The College of St. Benedict, USA

"This is a comprehensive, updated, indispensable textbook that addresses urgent religious and ethical issues raised by radical human enhancement biohacking technologies—for humanity today and in the future. This book should be translated for readers outside the English-speaking world."

—Heup Young Kim
Former Professor of Theology, Kangnam University
Director, Korea Forum for Science and Life

"This timely and thorough introduction to the many intersections between religion and emerging technologies provides readers from a variety of spiritual traditions with thoughtful, critical engagement with some of the most pressing technological issues of our time. Mercer and Trothen walk the reader through five categories of enhancements, Abrahamic and karmic religious traditions, and both supportive and critical religious responses to current and near-future enhancement technologies. Readers interested in the future of religion in a technologized society will welcome this volume for students and scholars alike."

—Amy Michelle DeBaets
Manager, Bioethics, Hackensack University Medical Center
Associate Professor, Hackensack Meridian School of Medicine

"The days of the computer Hal in Stanley Kubrick's *2001: A Space Odyssey* and of the androids in both versions of *Battlestar Galactica* have arrived. Computers, both hardware and software, have suffused our material culture, our bodies, and our minds. As the documentary *Social Dilemma* reveals, truth has been hijacked for the sake of profit. Machines and their information, for better or ill, generate and drive contemporary reality. The specter of machine manipulation of human awareness is no longer a Jungian dystopian archetype or metaphor. It has arrived. This important book provides students with the information and the discernment skills needed to negotiate this strange new world."

—Christopher Key Chapple
Doshi Professor of Indic and Comparative Theology
Loyola Marymount University, USA

"Mercer and Trothen tackle some of the thorniest and most exciting areas in the new era of science-religion dialogue—giving one of the most comprehensive overviews of the massive changes that are rapidly emerging, and the ways in which they will reshape our religious landscape. And they do this from a position of intimate familiarity with the movements, figures, and cultural forces that are leading this charge."

—Micah Redding
Founder and Executive Director, Christian Transhumanist Association
Host, Christian Transhumanist Podcast

"Momentous is the intersection of human enhancement technologies with religion. Mercer and Trothen are experts, and their textbook skillfully guides the student from awareness to understanding."

—Lincoln Cannon
Founder, Mormon Transhumanist Association

Calvin Mercer • Tracy J. Trothen

Religion and the Technological Future

An Introduction to Biohacking, Artificial Intelligence, and Transhumanism

palgrave
macmillan

Calvin Mercer
East Carolina University
Greenville, NC, USA

Tracy J. Trothen
School of Religion and the School of
Rehabilitation Therapy
Queen's University
Ontario, ON, Canada

ISBN 978-3-030-62358-6 ISBN 978-3-030-62359-3 (eBook)
https://doi.org/10.1007/978-3-030-62359-3

Cover Illustration: Sarah Holmlund / Shutterstock

This Palgrave Macmillan imprint is published by the registered company Springer Nature Switzerland AG.
The registered company address is: Gewerbestrasse 11, 6330 Cham, Switzerland

To Susan and Ron

Foreword

Welcome to your future! Opening this book is like opening a window to what lies ahead for you, your relationships, your world, and whatever you might think of as the underlying meaning of it all.

What you will find here is not some occult prophecy or fortune telling. There is much about the future that is simply unknowable. Not even careful and thoughtful scholars like Calvin Mercer and Tracy Trothen can foresee the twists or turns of politics or pandemics or the unpredictable shocks of "outrageous fortune," as Shakespeare's Hamlet so famously put it.

But this book is not just a bunch of wild guesses, either. It is based on a rigorous analysis of trendlines already in play, lines that point to the horizon of the future. We can see right now, for example, that technology is already changing humanity. We can see that its power to change us will surely grow in sophistication, precision, and scope. Follow these lines as far as your imagination can carry you. Where are they leading us? Where will they take you?

Barring something like an unexpected meeting with an asteroid, it is pretty obvious that human enhancement and technological transformation will continue. When that happens, we become enmeshed with our technology. We wear it, implant it, swallow it, and integrate it into our bodies and brains. We do not need to be "chipped" to be a cyborg. We are already merged with machines, byproducts of our own technology. We fuse the biological given with the technological unknown. And then we wonder why we are confused.

Our biohacked brains may still feel "natural" or "normal." In reality, however, there are no more "natural" humans, only technologized ones. Once upon a time we asked, "What does it mean to be human?" Now we ask whether any of the old definitions of humanity still offer any guidance. Keep in mind that the era of human enhancement is just getting started. Already we can see, however, that we are on our way toward becoming posthumans. We are former human beings who no longer remember what we once were but cannot imagine what we will soon become. Where will it all come out? I haven't a clue. But I am certain that we will be different from what we were in the past and what we are even now. We live suspended between old and new, on a zip line between the familiar and the unimaginable.

Welcome to the world of "trans-." We are all transforming or transitioning from one thing to another. We are reinventing ourselves without a blueprint or a goal. For some, human reinvention is scary. For others, it is liberating and exciting. For most of us, it brings mixed feelings. Here in this book, you will find reasons for all of the above. If technology is transforming humanity, is that good or bad, heaven or hell?

You will keep bumping into the question here, but you won't find a pre-set answer. You will be better informed about what is going on. You will have a clear, up-to-date base of knowledge about how technology is already at work to enhance human capacities or how it might develop in the future. You will have a comprehensive knowledge of the contemporary social debate over technology. You will meet the key players, the biohackers and philosophers and religious scholars who are driving today's arguments, ranging from the transhumanists and techno-enthusiasts to those who fear we are at risk for losing something old but really good about humanity. You will know what others think.

Knowing these things, however, won't determine how you respond. In fact, that's one of the things I really like about this book. It invites you to think but doesn't tell you what to think. It prepares you to engage the moment, to come to your own assessment, and to navigate your own personal pathway into the future.

In that sense, this is not a standard textbook, where everything is neatly packaged and settled. There is nothing neat or settled about the human future, yours personally or ours collectively. We have to live it to know it, and then we will not know fully even after it has happened to us.

No ... this is a textbook of a completely different kind, not full of yesterday's answers but uncompromising in pressing tomorrow's questions. It is a guide to what you might become. Welcome to your future self.

H. Parker Sharp Professor of Theology and Ethics, Pittsburgh Theological Seminary

Ron Cole-Turner

Preface

The idea for this textbook was born as your two authors dined in a restaurant on a busy street in one of the world's large cities, Boston. Having worked for years publishing technical journal articles and facilitating scholarly conversation on this topic, we were meeting with other scholars researching religion and radical human enhancement.

We watched, through the restaurant window, people busily walking, talking, eating, and conducting commerce. Most of our fellow human beings, we knew, had little or no understanding of the colossal impacts on their lives that are coming as a result of what we and our handful of academics around the world were debating among ourselves. That busy street scene could look discernibly different in just a few years, in ways that most people could not imagine.

At that window on a piece of our busy world, we decided it was time to widen the conversation, to do our part to share with students and the general public some of the profound changes aborning, and specifically how those changes might intersect with religion. This textbook is our attempt to engage you—and engage you profoundly—in the conversation about radical human enhancement, a conversation topic that will increasingly take its place on the mainstage of our lives politically, economically, socially, religiously, and all other ways.

We completed this textbook amidst the COVID-19 pandemic, a health and economic event that is altering the world from the time of our Boston musings. Epidemiologists, virologists, engineers, medical professionals, and many others have been coming together to save lives. The pandemic brought into relief the interconnectedness of life and the remarkable capacity of humans to do good through the application of science. We also witnessed greed and protectionism. The pandemic made us even more acutely aware of the many ways human potential for both good and evil can be amplified by technology. We trust our collective experiences of the pandemic will inspire all of us to take more seriously the topic of radical human enhancement and the critical role of religion.

Greenville, NC, USA Calvin Mercer
Kingston, ON, Canada Tracy J. Trothen
January 1, 2021

Acknowledgments

Knowledge and inspiration flow from teachers to students—and vice versa. Calvin thanks his students, especially in his "Religion and Science" classes and his advanced seminar on religion and radical human enhancement, for their insights and passion. Thanks goes to Dr. Alan Schreier, resource person for the classes, for providing science and technology expertise as we pursue interesting and important questions with students—and for keeping Calvin honest on the science side. Dr. Schreier is a MIT-trained biochemist with an interest in extreme human enhancement. A number of family members read an earlier draft and provided encouragement and stimulating conversations.

Some brief sections of this book are adapted from the introductory chapters of two of Calvin's coedited books: *Religion and the Implications of Radical Life Extension* and *The Body in the World's Religions: Reflections in Light of Transhumanism*, both published by Palgrave Macmillan.

Tracy, too, is grateful to her students and to colleagues and friends who have shared an interest in what it means to make us better. She is also grateful for the opportunity to discuss technology, AI, and ontological questions at the American Academy of Religion (AAR) with diverse, wonderful colleagues, including Randy Reed who did all the work but graciously invited her to co-found and co-chair the AAR's recurring Artificial Intelligence Seminar.

Previous publications of Tracy's that have been helpful in writing this book include: *Spirituality, Sport, and Doping: More than Just a Game*; two concluding chapters from books she has coedited with Calvin, namely *Religion and Human Enhancement: Death, Values, and Morality* and *Religion and Transhumanism: The Unknown Future of Human Enhancement Technology*; and her chapter in *Technology and the Image of God: A Canadian Conversation*, published by The Canadian Council of Churches.

Acknowledgments

Contents

About the Authors

Calvin Mercer, PhD Mercer was trained as a biblical scholar and published in that field for years. For a decade and a half, he has addressed the opportunities, dangers, and religious implications of radical human enhancement technology. Along with UK scholar Steve Fuller, Mercer coedits *Palgrave Studies in the Future of Humanity and Its Successors*, the oldest book series addressing human enhancement from all disciplines and points of view. Mercer's publications on human enhancement are listed in the bibliography. He was founding chair of the "Human Enhancement and Transhumanism" Unit in the American Academy of Religion, the largest and most significant organization devoted to the academic study of religion, and he currently serves as co-chair of that unit's steering committee. He serves on the Academic Advisory Council of the Christian Transhumanist Association. Also trained in clinical psychology, Mercer practiced professionally part-time for over a decade and uses this discipline in his published work on religion. He is professor of religion at East Carolina University and gives public lectures on the religious implications of human enhancement technology, fundamentalism, and other topics.

Tracy J. Trothen, ThD Trothen is Professor of Ethics at Queen's University, jointly appointed to the School of Religion and the School of Rehabilitation Therapy, where she teaches in the graduate "Aging and Health Program." She is a certified Supervisor-Educator in Clinical Spiritual Health (CASC), Registered Psychotherapist (CRPO), and ordained minister in The United Church of Canada. Her numerous publications on human enhancement are listed in the bibliography and include *Spirituality, Sport, and Doping: More than Just a Game*, and *Winning the Race? Religion, Hope, and Reshaping the Sport Enhancement Debate*. Her areas of research and teaching specializations include biomedical and social ethics, moral distress, technology, and religion and sport. Trothen is on the steering committee of the American Academy of Religion's "Human Enhancement and Transhumanism" Unit, and she is a founding co-chair of the Artificial Intelligence Seminar.

PART I

Setting the Scene

CHAPTER 1

Introduction

The End of Religion—Or Not

The religions of the world will come to an end, or thrive, depending on how they respond to the topic of this book. You may judge that an exaggerated statement. That is fair enough. Indeed, it is a drastic statement, we grant that. But we do not think it is exaggerated or untrue. We invite you to consider the information provided and the issues raised in the following chapters before drawing your own conclusion.

There is a new world coming, and it is coming much faster than most people realize. In the following chapters, we describe the various categories of radical—some might prefer the word "extreme"—human enhancement and the biohacking technologies facilitating them. You will learn about the world that will be faced by your grandchildren, likely your children and, to some degree, you. Elements of this new world are already present; others are just around the corner.

The term "biohacking" is used in different ways. We use it to mean the radical enhancement of human beings with therapies and technologies now available or anticipated in the future. For good, evil, or in-between, our scientific knowledge and technological innovations are growing rapidly, and we can envisage an elevated tempo in the coming decades.

Now we have Netflix, live streaming, wide-ranging plastic surgeries, pacemakers, self-driving cars, smart homes, drones, and more. The technological field is exploding, with enormous implications for biohacking ourselves. Artificial intelligence adds significantly to the ability to enhance ourselves, to make ourselves better. What it means to become better and who gets to decide are key ethical questions explored in the coming chapters.

It is a moral imperative that we who are alive now take steps to ensure the continuing transformations unfold responsibly. Religion can have much to say in the public square about efforts to biohack our way to better. Faith

C. Mercer, T. J. Trothen, *Religion and the Technological Future*,
https://doi.org/10.1007/978-3-030-62359-3_1

communities, religion scholars, and students engaged in the study of religion can play important roles in advocating for the responsible use of human enhancement technology, if we create space for dialogue and reflection. We intend this book to help to create that space.

The relationship between religion and human enhancement technologies cuts two ways: (1) what religions say about the enhancement technologies, and (2) what impact the technologies have on the religions and their belief claims. We explore these reciprocal questions as we discuss each human enhancement category.

While our primary focus is on religion, inevitably radical enhancements are going to impact every human institution. If superlongevity, just to pick one enhancement, is realized and the technologies become widely available, it would arguably have a more radical impact than any other development in human history. One need only reflect briefly on the economic, political, governmental, medical, family life, and social implications of people living for such long periods to realize the significance of such a development. The discussion of the religious implications of radical enhancement can provide general indicators of impacts on these other domains of human experience.

The Chapters That Follow

In each chapter, we describe the enhancement technology and explain how it is upon us, just around the corner, or farther into the future. We reflect on the religious implications of the technological enhancements. Our twofold focus is, first, on belief, often more technically referred to as theology or doctrine, and, second, on the ethics of biohacking our way to being better. For followers of the faith traditions, one's theology, ideally, informs their ethics and, therefore, their behavior in the world.

Radical human enhancements will impact not only theology and ethics. The technologies will also impact religious rituals, institutions, spirituality, and all other facets of religion. While we occasionally make illustrative reference to a concrete ritual or institutional structure, our focus is on the broader theological and ethical implications of the enhancements. Conversations about these topics by scholars of religion are underway, but they are not developing nearly as quickly as are things on the technological side.

Academics in all disciplines, as well as the general public, are still in the very early stages of understanding, much less responding to the fast descending enhancement technologies. While all religions are going to be reshaped, if they survive at all, at this point the most extensive scholarly conversations about human enhancement and religion are unfolding in the two religions of Judaism and Christianity, and even these conversations are only at the beginning stages of addressing the issue. By necessity, then, most of our examples are from these two traditions. However, we understand the need to engage the pressing questions of radical enhancement from a diversity of religious perspectives, and we will work to begin that process.

With regard to the technologies and research trends addressed in the following chapters, some are in the theoretical or experimental stages. Many, however, are in place and constantly being improved. In some cases, the technologies are developing so quickly that they may have changed significantly by the time this book is in your hand or on your screen. Our goal is not to report in detail the latest technological advances, although we include examples of cutting-edge research. Rather, our goal is to depict the basic thrust of the relevant human enhancement technologies in order to show what the religions are facing and how they are beginning to respond to the challenges.

The textbook is divided into three main parts. The first part, "Setting the Scene," consists of four chapters. Following this introductory chapter, in Chap. 2, "Existing and Possible Technologies: How We Biohack," we explain, in general, what we mean by radical human enhancement and show how it is already occurring. Then, without become too technical, we summarize some of the major therapies and technologies that are and likely will be employed for enhancement purposes. These summaries establish a basic knowledge of technologies that are referred to throughout the textbook.

Chapter 3, "Transhuman, the Posthuman, and the Religions: Exploring Basic Concepts," defines the terms "transhuman" and "posthuman" and provides a brief history of transhumanism as an intellectual and cultural movement. Religion is, of course, expressed concretely in the religions of the world, so we outline major theological themes of the world's faith traditions that can be useful in interpreting human enhancement technology. We distinguish some basic differences between the monotheistic and karmic religions that can inform how traditions originating in different parts of the world might react to the radical enhancements. We do not provide an exhaustive exploration of the religions; our intention is to spur a more robust conversation about the traditional religions and other faith-inspired paths with regard to radical enhancement.

The final chapter in Part I is titled "Radical Human Enhancement and Ethics: Questions We Must Ask" and introduces the discipline of applied ethics. The moral relevance of context, systemic power imbalances, and perspective are considered. Consultation, as a way of involving diverse voices in ethical decision-making, is emphasized. Readers have opportunity to identify their values, as a way of illustrating the role of values in making ethical choices. After introducing some ethical theories, three ways of defining the human enhancement issue are presented: (1) the therapy–enhancement continuum and ways to make us better, (2) choice, and (3) justice. Each of these three ways is considered through the application of a religiously informed lens. Finally, precautionary and proactionary approaches to radical human enhancement are explained.

Part II, "Five Categories of Enhancement," consists of three chapters examining the five categories of radical human enhancement. In each chapter, a variety of religious and ethical issues are addressed that were introduced in Part I of the textbook. In Chap. 5, "Superlongevity and Other Physical

Enhancements," we explore physical enhancement and, particularly, the termination of aging. Chapter 6 covers two enhancement categories and is titled "Cognitive Enhancement and Moral Bioenhancement: Becoming Smarter and More Moral." Affective enhancement and spiritual enhancement are explored in Chap. 7, which is titled "Affective Enhancement and Spiritual Enhancement: Feeling Happier and More Spiritual."

The third major part, "Special Topics: Going Beyond the Edge," takes up three topics that have made their way into the human enhancement conversation. "Cryonics: Buried, Burned, or … Frozen," Chap. 8, explains cryonics, which some transhumanists see as a stopgap measure, keeping people alive in a "frozen" state until technology is invented that gives them the choice to live indefinitely. Chapter 9, "Mind Uploading: Cyber Beings and Digital Immortality" introduces what is more technically called whole brain emulation, that is, transferring the memory and personality into a more reliable platform than human biology. Finally, Chap. 10, "Superintelligence: Bringing on The Singularity" explores "AI on Steroids," a development that could bring untold benefits to humanity and the planet or end life as we know it. In all three chapters in this third part of the textbook, a variety of religious and ethical issues are probed.

The textbook concludes with Chap. 11, "Religion 2.0 and the Enhanced Technological Future." Many themes from prior chapters are woven together, and new themes introduced, to show how the religions might chart a path forward to effectively address radical human enhancement technology.

Your Authors and Our Goal

Your authors have been studying, teaching, and writing about radical human enhancement for well over a decade and a half. Our backgrounds are in the academic study of religion and ethics. While Professor Mercer brings broad knowledge of the world's religions and Professor Trothen has expertise in ethics, we have jointly processed each section of the textbook. Also, we have vetted the technical side of the following chapters with those who can keep us honest on that front, although we are quick to say any errors that slipped through on the technical front are ours alone.

We put forward plenty of ideas, but please do not read our reflections as mandating what religious responses should be. Our goal is to introduce religious and ethical discussion points in the interest of fostering dialogue and learning. Our commitment is to stimulate informed conversation about human enhancement technologies and their implications. All reasonably argued points of view are encouraged. That said, we will certainly suggest enough positions to spark discussion and, hopefully, productive disagreement.

Scholars involved in the academic sessions we have helped organized at the annual meetings of the American Academy of Religion (AAR), and those who have contributed to the books we have coedited, include both strong proponents and sharp critics of radical human enhancement. We designed it that way.

We are strong advocates of the free exchange of ideas and vigorous debate as religious, theological, ethical, and policy positions get hammered out. To assist you in digging deeper, we have included questions for discussion at the end of each chapter. A glossary of important words, as we are using them, is found at the end of the textbook.

Who Are Our Readers?

Some readers of this textbook will identify not only as a student of religion but as a follower of a religion or as spiritual, more generally. For you, the textbook is an opportunity to reflect on technological developments, including artificial intelligence, that will increasingly impact your faith tradition or spirituality. Some of the extreme enhancements being discussed, such as uploading the brain's information to a computer platform, may not materialize. Some of them, such as altering mood with pharmaceuticals, are already happening. It is incumbent on adherents of all faith traditions to reflect theologically and ethically on these developments and to weigh in on the needed public conversation. Once these issues make their way further into our political debates—and they will—it is important that religions be well represented in the public square and that religiously informed concerns are included in policy construction and in the creation of new tech.

Readers who are not religious but who are interested in the place of religion in society are introduced to the topic of biohacking to achieve radical human enhancement and how these developments are impacting and may impact religion. Religion, for better or worse, has played important roles in the history of societies around the globe and is integral to understanding political and cultural events. It is important for informed people, even if not religious themselves, to understand the coming upheavals in religion and society that will accompany radical human enhancements.

Some readers of this book have more facility with the science and technology fueling radical human enhancements than other readers. Perhaps you are majoring or concentrating in one of the sciences or in some technical field, such as computer science or genetics. If so, then you may have some background for understanding the relevant human physiology, or informational science, of some of the technologies discussed. Even so, you may not have realized the full implications for human enhancement of your scientific or technical field. Other readers may have a stronger background in one of the liberal arts, such as religion, history, or literature. We have presented the technical sections of the chapters in ways that can be useful for those with diverse academic backgrounds.

Before We Continue, Here Is a Survey

We invite you (and perhaps your class or study group) to consider how you describe yourself with regard to radical human enhancement, at this moment, before you move through this textbook. With regard to radical biohacking, are you:

- Totally skeptical and opposed.
- Open to the possibilities but needing to be convinced about each enhancement.
- An eager transhumanist, i.e., an eager advocate for radical enhancement.
- Unsure.

Keep a record of your response. At the end of our study, we give opportunity for you to answer the same questions and compare your responses, reflecting on how they might have changed and why.

CHAPTER 2

Existing and Possible Technologies: How We Biohack

What We Mean by Radical Human Enhancement

A growing group of scholars and thought leaders believe that biomedical, machine, and other technologies will increasingly refigure the human condition. Not that long ago, some of the scientific advances discussed in this textbook were the stuff of science fiction. No more. Scientific developments that may extend our lives and improve aspects of our being are unfolding in university, government, hospital, and corporate labs around the world.

A few journalists have been reporting this story, and a few scientists, ethicists, and authors are debating it. The media carries stories about relevant medical breakthroughs and occasionally suggests their more radical implications. Artificial intelligence (AI) in particular is capturing attention in the popular media. However, the significance of these developments has not yet dawned in the public's mind, despite a few cover stories in news magazines and TV specials.[1]

We already commonly replace knees, hips, fingers, elbows, and shoulder joints, and we have cochlear implants and electronic devices to improve or restore hearing. Prosthetic devices are available for hands, arms, and legs. Bionic hands are being improved. Researchers in Japan and elsewhere are developing 3-D printers that combine stem cells and artificial materials to custom make artificial ears.

Radical enhancements are likely to be in high demand, especially for younger generations who often are more comfortable with new ideas and technologies. The interest in biohacking our bodies and minds is evident from the popularity of psychotropic and sexual performance pharmaceuticals, cosmetic surgery, and sports medicine. These successful efforts to impact our mood and body

[1] E.g., Harry MacCracken and Lev Grossman, "Can Google Solve Death?" *Time* (September 30, 2013). http://content.time.com/time/covers/0166412013093000.html.

C. Mercer, T. J. Trothen, *Religion and the Technological Future*, https://doi.org/10.1007/978-3-030-62359-3_2

pale against the wave of engineering technologies on the horizon. Existing and emerging technologies give new meaning to an old religious passage, "The blind receive their sight, the lame walk, the lepers are cleansed, the deaf hear, the dead are raised ..."[2]

A thoughtful theological and ethical assessment of technology and its role is critical for today's world religions and, indeed, all of society. Technologies or medical procedures initially seen as shocking and pushing the bounds of acceptability often become seen as normal over time. Heart surgery is a good example. In 1944, surgeons at Johns Hopkins Hospital carried out the first successful heart surgery on an infant with Tetralogy of Fallout, a rare congenital condition. This surgery shocked the world, and some questioned whether we should be intervening in God's domain of the human heart. Now, cardiac surgery is regarded as a normal therapy. We will discuss the meaning of "normal" and how it relates to human enhancement ethics in the chapter, "Radical Human Enhancement and Ethics."

We Are Already Enhanced

Professor Mercer, one of your authors, is writing his draft of this section while being transported in a very high-tech machine, an airplane, piloted by a computer. He is writing the draft on a computer, occasionally glancing through technology resting on his nose (i.e., eyeglasses) and at the time machine on his wrist to determine how much longer he has to work before landing.

Technology is essential to your reading these sentences right now, whether you are doing it on traditional codex book technology, a Kindle, or the newest iPad or iPhone. Technology is literally all around and all over us. We could add "and in us." We speak not just about artificial knees and hips. Even as you read this book, chips—"neural implants" is the more accurate and technical term for these computer bits—are being implanted in people's brains to treat strokes or to further scientific research. Welcome to the new world of neurotechnology and biohacking. These and other enhancements are becoming more common and potent.

We can observe the trajectory of technology by considering how we communicate to each other. Baby Boomers and some Gen-Xers remember when telephones sat on a table beside the recliner, with a cord running into the wall and outside to the telephone line. Then came the day when you walked around the house and spoke to your friend on a wireless phone. The phone device eventually moved to a belt on the hip, and now it is common to see people talking via a device hooded to the ear. Phone communication technology has moved from being outside on a table to being attached to the body. Technologies

[2] The New Testament, Matthew 11:5. All Bible references are from the *New Revised Standard Version*.

enhancing communication will not stop at devices attached to the outside of the ear. They will move into our bodies as implants.[3]

Short for "cybernetic organism," a cyborg is an organism integrated with artificial technology whose operation restores the organism's so-called normal ability or takes it beyond normal functioning. Cyborgs abound in science fiction. But this machine technology is not limited to science fiction imaginings. We are already primitive cyborgs. Even the addition of eyeglasses is a human interface with technology. Most everyone knows someone with implanted pins that have been used to reconnect bones or fasten artificial joints.

Pacemakers reflect the path of much of our enhancement technology. They were worn externally at first and then, in 1958, the first one was implanted. The first patient died in 2001, outliving the surgeon who implanted it. The technology of pacemakers, now routine, continues to become more sophisticated. The newer pacemakers are smaller, flatter, and last longer. Some are Wi-Fi connected. We could chart a similar history for cochlear implants. Or, we could turn to vision medicine and note the dramatic shift from contact lens to the cutting-edge work on bionic eye-brain implants.

Enhanced Athletes

To highlight yet another area, consider the impact pharmaceuticals and other technologies are having on sports and the intense ethical debate surrounding that. One of your authors, Professor Trothen, is a leader in addressing sport enhancement ethics. While most people are quite certain that anabolic steroids do not belong in sports, the general public is much less certain about the use of other enhancements, such as Tommy-John surgery and Smart Bats.

In 1974, professional baseball pitcher Tommy John was the first to undergo surgery to repair damage done to his ulnar collateral ligament, by using tendons from other areas of the body. Some think the surgery actually improves one's ability to throw, although medical experts disagree. But the belief that the surgery is enhancing has led other pitchers, and aspiring pitchers' parents, to request the surgery, even if there is no injury.

Many common-place enhancements in elite sport have not yet made the headlines and become well-known. Artificial intelligence has made its way into sport. Smart Bats in baseball enhance the batter's swing through real-time feedback on batter movements. In football, robots take hard—possibly concussion-inducing—tackles. Brain stimulation techniques stave off exhaustion in endurance athletes when transcranial direct current stimulation (tDCS) changes the perception of effort, making it more possible to push through exhaustion and pain.

Pharmaceutical central nervous system stimulants, such as methylphenidates (e.g., Ritalin), increase concentration and focus. Almost one in ten professional

[3] E.g., Lev Grossman and Matt Vella, "How Apple is Invading our Bodies," *Time* (September 10, 2014). http://time.com/3318655/apple-watch-2/.

baseball players are prescribed Ritalin for attention deficit disorders. The concern is that only four to six percent of the general population are prescribed this drug. Because elite sport is all about winning, and athletes win by the smallest of margins, anything that might give a boost, no matter how small, is highly sought. As a result, enhancement technology often shows up first in the elite sport domain. The big money in professional sport makes acquisition of expensive, powerful enhancements more possible.

Technologies and Research Trends

The following human enhancement technology trends and research programs illustrate the broad scope of unfolding technologies that will impact the future of the human species. This list is not exhaustive but does give a sampling of what is emerging. We identify more interventions throughout the textbook. Some of the areas of technology overlap.

Genetic Engineering and Therapy

Scientists are working to identify genes responsible for aging and various diseases and add, delete, or alter genetic material at specific locations in the genome to improve health and possibly extend life. A couple of examples illustrate this exploding field. In 2017 Kymriah became the first gene therapy to receive Food and Drug Administration (FDA) approval in the United States. Within three months of treatment with Kymriah, a dramatic 83 percent of all patients had either complete remission of their deadly form of leukemia or complete remission with incomplete blood count recovery.[4] Philadelphia-based Spark Therapeutics now has Luxturna, an eye gene therapy, on the market as the first FDA approved gene therapy for an inherited disease. Both Kymriah and Luxturna use an altered adeno-associated viral vector to deliver a healthy gene to replace a mutation in the RPE65 gene that causes inherited degenerative retinal disease.

A transformative breakthrough is CRISPR (clustered regularly interspaced short palindromic repeats), a technology being developed at highly reputable places, such as MIT and Harvard University. CRISPR can locate genetic code in DNA, and then an enzyme that is produced by the CRISPR system, such as Cas9, binds to the DNA and cuts it when the target DNA is located. Cas9 stands for "CRISPR associated protein 9." Cas9 is an enzyme that acts like a pair of scissors to cut the two DNA strands in specific places in the genome to allow for the addition or deletion of DNA material. CRISPR-Cas9 allows for the precise, quick, and inexpensive editing of DNA. Other systems in addition

[4] E. Seger, "FDA Approves First Gene Therapy for Leukemia Treatment: The How's, Why's, Promise, and Peril," *The Science Distillery* (September 5, 2017). https://sciencedistillery.com/2017/09/05/fda-approves-first-gene-therapy-for-leukemia-treatment-the-hows-whys-promise-and-peril/.

to Cas9, such as CPF1, are being developed. Cas9 can be modified to activate genes as well as delete them. Gene editing technologies such as CRISPR-Cas9 may allow for senescent (growing old) cells to regain youthful structure and function. Much depends on our ability to identify relevant genes and clusters of genes. Only time will tell how this science proceeds, but there is much optimism.

George Church, of Harvard Medical School, in a remarkable statement, said, "A scenario is, everyone takes gene therapy—not just curing rare diseases like cystic fibrosis, but diseases that everyone has, like aging."[5] CRISPR is not some fringe mad scientist idea. Church's company, an IPO, is backed by Google Ventures and the Bill and Melinda Gates Foundation.[6]

Even before CRISPR technology, major increases in lifespan had been achieved in several animal species through genetic intervention. For example, changes in just one gene of the nematode worm can significantly extend the worm's life. In some male worms, the lifespan has increased six-fold. Similar research has been conducted on yeast, fruit flies, and mice.[7] The goal of these scientific inquiries, of course, is radical human life extension. Stem cell research is expected to enhance the development of genetic manipulation technologies. Completion of the Human Genome Project in 2003 may one day be seen as the turning point between understanding human biology and engineering it.

Germline Genetic Engineering

Somatic genetic engineering, discussed above, impacts only the individual who receives the treatment, not their descendants. CRISPR could be harnessed to ultimately push genomic changes through generations or even through an entire species. This potential for germline genetic engineering, as opposed to somatic genetic engineering, disturbs many.

Germline genetic engineering refers to manipulating "germ" cells (i.e., the egg, sperm, and embryo cells). Engineering the genetic material in a human embryo means that any changes made are passed down generation to generation. Once this threshold is crossed, it may be the first big step toward the end of humanity as we know it. This capability would allow for the suppression and

[5] Joel Achenbach, "A Harvard Professor Says He Can Cure Aging, But Is That a Good Idea?" *Washington Post* (December 2, 2015). https://www.washingtonpost.com/news/achenblog/wp/2015/12/02/\professor-george-church-says-he-can-reverse-the-aging-process/.

[6] In February 2020, a US trial used CRISPR-Cas9 gene editing, with no side effects, on the immune systems of three cancer patients. (AFP, "US Trial Shows 3 Cancer Patients Had Their Genomes Altered Safely by CRISPR," *ScienceAlert*. Accessed February 9, 2020.)

[7] Mark Geanacopoulos, "The Determinants of Lifespan in the Nematode Caenorhabditis Elegans: A Short Primer," *Science Progress* 87, no. 4 (2004): 227–247; Anders Olsen, Maithili C. Vantipalli, and Gordon J. Lithgow, "Using Caenorhabditis Elegans as a Model for Aging and Age-Related Diseases," *Annals of the New York Academy of Sciences* 1067, no. 1 (June 2006): 120–128. https://doi.org/10.1196/annals.1354.015; and Nitish Mittal, et al., "The Gcn4 Transcription Factor Reduces Protein Synthesis Capacity and Extends Yeast Lifespan," *Nature Communications* 8, no. 1 (2017). https://doi.org/10.1038/s41467-017-00539-y.

enhancement of naturally occurring conditions and traits that would then be passed on to future generations.[8] It could become possible to design out obesity, poor eyesight, and cancer, while designing in tallness, musical ability, higher IQ, and longevity. Genetic engineering also, understandably, brings up concerns about how these potent technologies could be used for nefarious ends. Germline editing raises additional ethics questions around the consent of future generations that we will explore later in this textbook.

Germline genetic engineering may bring with it currently unknown side-effects, including off-target side effects, which are unintended impacts on genes other than those being targeted for change. So, while there are still many technical difficulties to overcome before "designer baby" engineering could be routinely and safely utilized with humans, CRISPR-Cas9 moves us closer to this possibility, and that disturbs many. Already, in China, twins have been born whose embryos had been edited by CRISPR technology to prevent the possibility of HIV. Since the edited DNA can be passed to their progeny, and because of the dodging of regulatory accountability, much world-wide condemnation of this unapproved gene-editing experiment arose.

Gene drives could push genomic changes through an entire species. A gene drive is genetic engineering technology that can be used to change genes in germline cells to propagate changes through an entire species. This engineering technique changes genes so they do not keep to the typical hereditary patterns. Unlike germline genetic modification technologies, this technique ensures that the new gene gets passed on to all generations, indefinitely. Typically, genes have a 50 percent chance of being passed on. Gene drives move that percentage to 100 percent. The amount of time that it would take to change an entire species depends on the length of the species reproduction cycle.

Mosquitos, as a species, could be genetically altered to prevent the spread of malaria. Because of their relatively short reproduction cycle, the entire species could be changed in about one year. While the eradication of malaria spread via mosquitos may sound very attractive, we do not know all the implications and off-target effects of this change. Gene drives, and germline genetic engineering, are important emerging biotechnologies that raise many theological and ethical questions for the religious traditions.

Tissue Engineering and Organ Replacement

Regenerative medicine, a broad field that includes tissue engineering, is pushing forward on repairing, improving, and replacing damaged or diseased organs and tissues in our bodies. We now routinely replace knees, hips, and kidneys (the latter through organ transplant, often from living donors), and indications are that these body part replacement procedures will become more common.

[8] K. E. Ormond, et al., "Human Germline Genome Editing," *American Journal of Human Genetics* 101, no. 2 (August 3, 2017): 167–176.

Certainly, the brain and nervous system are in a different league; however, there are research programs that may impact replacement of these significant parts or facilitate their regrowth.

Research on therapeutic cloning, that is, producing organs and tissues for transplanting into humans is proceeding, despite ethical debate. At the pioneering Wake Forest Institute for Regenerative Medicine,[9] in Winston-Salem, North Carolina, USA, cells and/or tissues from skin, urethras, cartilage, bladders, muscle, kidney, and vaginas have been grown in the lab and implanted into human beings. The Institute was the first in the world to implant a laboratory-grown organ into humans. Using tissue engineering, cell therapies, biomanufacturing, nanotechnology, gene editing, and 3D printing, the research teams there are working on growing more than 40 different tissues or organs. Regenerative medicine has been called the "next evolution of medical treatments," by the United States Department of Health and Human Services.[10]

Theoretically, we could replace many more of the parts of our bodies than are currently possible. When the transmission in your car breaks, you replace it and you still consider it your car. What if, over time, you replaced every part of your car, would it be the same car? This is a modern version of the ancient "Ship of Theseus" thought experiment discussed by Greek philosophers. As the tissue engineering and organ replacement possibilities expand, theological and ethical issues having to do with personal identity and human nature press upon us.

Artificial Intelligence

When we think of AI, we might recall Hollywood movie depictions of death-dealing robots and androids battling with us to take over the planet. But the AI image is changing as we hear about Deep Blue, the masterful computer that beat the best humans at chess over 20 years ago, or the more recent AlphaGo computer that excelled at "Go," perhaps the most complex boardgame ever devised. Now, we have AlphaZero and Stockfish 8 as renowned AI chess masters. AI is attracting, appropriately, much attention with regard to its impact on jobs and the economy. With self-driving vehicles and smart home devices, such as Siri and Alexa, AI is increasingly becoming a part of our daily lives. Having moved through a number of boom and bust periods, AI is positioned for significant advancement.

This "weak" or "narrow AI" involves computers programmed to perform a particular function with a predefined range and provides the context for the development of "strong AI, also called Artificial General Intelligence (AGI)," which is machine intelligence that mimics that of humans. At this point theoretical, strong AI is anticipated by many experts to be developed in the coming

[9] Wake Forest Institute for Regenerative Medicine. https://school.wakehealth.edu/Research/Institutes-and-Centers/Wake-Forest-Institute-for-Regenerative-Medicine.

[10] Ibid.

decades. Weak AI is neither conscious nor sentient. Strong AI, however, moves us into capabilities resembling human intelligence and all the theological, ethical, and other attendant issues. We discuss strong AI and what might come beyond it—superintelligence—in the chapter, "Superintelligence."

Merging of Computer Technology with Human Biology

Pacemakers and a variety of other implants are now commonplace, but someday computers that interface with the human body may become even more routine. Biomechatronics, pushing forward the intricate integration of biology, mechanics, and electronics, is a growing research field and is related to the also exploding fields of robotics and neuroscience. We have implants to help deaf people hear, and a retinal implant helps some blind people see. Neural implants are placed directly into the brain of some patients to counteract symptoms of Parkinson's Disease and other neurological disorders. Increasingly, computer technology is being embedded in our bodies.

With research expertise in AI, robotics, and biomedical engineering, Emeritus Professor Kevin Warwick at Coventry and Reading Universities in the UK is famous for his pioneering experiments involving the implantation of technology into the median nerves of his arm. This device allowed him to link his nervous system to a computer, helping research move forward that might someday assist people with disabilities limiting their ability to communicate.[11]

Robotics

Robots are being used more and more in various areas of life. A team of undergraduate engineering students at Dartmouth College developed the Mobile Virtual Player (MVP), a robot that can stand in for players and take hard hits that may otherwise cause concussions during football practice sessions. These robots are automated by remote control and engage with players at the direction of the coach operating the controls. More autonomous robots that could participate in actual plays in real-time are planned for the future. Life-sized sex dolls have been around for years. Artificially intelligent robots are now being marketed and utilized for companionship, beyond mere physical sex. All of these AI creations, in addition to biohacking processes, are designed for human enhancement.

Utilizing advanced tissue engineering capabilities, robots of the future will not necessarily be comprised of only plastic and metal. These machines are becoming more and more humanoid, bringing on reports about the "dehumanization of robots," even though they are not human.[12] While we usually think of robots as machines, separate from us, robots can play an important

[11] www.kevinwarwick.com.

[12] Jonah Englel Bromwich, "Maybe We Are Wired to Beat Up Machines," *The New York Times, SundayStyles* (January 20, 2019): 1, 8.

role in enhancing our physical abilities, especially when designed so as to smoothly integrate with flesh and blood bodies.[13] Artificial intelligence will get ever more sophisticated, allowing robots to be (at least on the surface) sensual, social, and emotional, and in the process raising theological and ethical issues.

Other Technologies

A variety of other cutting-edge technologies can be marshalled in the service of human enhancement. Extended reality (XR) is an umbrella term that includes various forms of simulated environments. Virtual humans and digital immortality have been under development for years.[14] Commercial competition, for example, in the virtual reality video game industry, is just one factor fueling investments in the development of such technologies.

Brain stimulation techniques are used to treat major depression and aid in recovery after strokes. They can improve athletic performance, and rumors are they have been used by athletes in Olympic games. Transcranial direct current stimulation (tDCS) improves endurance. Even though not yet banned in sport, tDCS does have strong potential downsides. It can cause seizures, headaches, and changes in thought patterns. Deep brain stimulation utilization is emerging all over the world.[15] A quick Google search finds brain stimulation devices on Amazon—even travel sized—for about 260 Canadian dollars.

Non-invasive scanning techniques, such as the MRI, PET, and CAT, are routinely employed. Brain scanning technologies are fast increasing in resolution. Conceivably, scientists will eventually peer inside synapses and record neurotransmitter activity, obtaining detailed models and simulations of all regions of the brain. If the brain is completely mapped, the job of redesigning and rebuilding the brain could be furthered considerably.

Nanotechnology is an exciting, emerging science that almost assuredly will play a role in human enhancement. A "nano" is one billionth of a meter, the width of about five carbon atoms. Nanotechnology research is active with animals and may eventually produce blood-cell sized, computerized tools called nanobots, capable of manipulating human biology at the cellular level.

We have only touched on the myriad of technologies that are being used or being considered for human enhancement. Whether or not they all make us

[13] For an accessible introduction to robotics by a leading expert, see Rodney A. Brooks, *Flesh and Machines: How Robots Will Change Us* (New York: Vintage Books, 2002).

[14] E.g., Martine Rothblatt, *Virtually Human: The Promise—and Peril—of Digital Immortality* (New York: St. Martin's Press, 2014). Rothblatt is a lawyer, medical ethicist, technologist, and entrepreneur. She addresses the technical, legal, moral, and even spiritual questions surrounding the development of "mindclones," digital copies of our identities. She believes religions will eventually embrace mindclones. See also N. Katherine Hayles, *How We Became Posthuman: Virtual Bodies in Cybernetics, Literature, and Informatics* (Chicago: University of Chicago Press, 1999); and Debra Bassett, *The Creation and Inheritance of Digital Afterlives: You Only Live Twice*, in *Palgrave Studies in the Future of Humanity and Its Successors*, series eds. Calvin Mercer and Steve Fuller (New York: Palgrave Macmillan, forthcoming).

[15] Chencheng Zhang, et al., "An International Survey of Deep Brain Stimulation Utilization in Asia and Oceania: The DBS Think Tank East," *Frontiers in Human Neuroscience* 06 (July 2020). https://www.frontiersin.org/articles/10.3389/fnhum.2020.00162/full.

better is not a simple question allowing for an easy answer. There are many issues to be considered. Religion, as we will see, helps frame the complexities of emerging technologies and of being human.

Questions for Discussion

1. Most of us use a smart phone and computer. Are we already being (unconsciously?) drawn into a world of significant human enhancement? Is this a good thing?
2. How would you distinguish the line between "normal" and "radical" enhancement? What are some examples of "normal," as opposed to "radical," enhancements?
3. What issues of fairness are involved when athletes use enhancement technologies and pharmaceuticals to improve performance?
4. Why might somatic genetic engineering be acceptable and germline genetic engineering not be acceptable?
5. Identify some of the ways AI, including smart robots, is at work in your daily life.
6. Identify some of the technologies that might work in concert with each other to produce even more potent enhancements.
7. Were you aware of the enhancing technologies discussed in this chapter? If not, why do you think you had not heard of them before?

CHAPTER 3

Transhumanism, the Posthuman, and the Religions: Exploring Basic Concepts

What Is Transhumanism? What Is the Posthuman?

The term transhumanism is usually traced back to an often-quoted statement by evolutionary biologist and philosopher, Julian Huxley (1887–1975), who wrote in 1957:

> The human species can, if it wishes, transcend itself—not just sporadically, an individual here in one way, an individual there in another way, but in its entirety, as humanity.[1]

The history of the term is more complicated, and interesting, than that, however. Natasha Vita-More provides a detailed etymology of "transhumanism" and similar terms, revealing their usages by the Italian poet Dante (c. 1265–1321), T. S. Elliot, FM-2030, and others. Vita-More credits FM-2030 (f/k/a FM Esfandiary) with first using transhuman as an evolutionary concept in the 1970s.[2]

Professor Ron Cole-Turner has a long and distinguished career speaking thoughtfully about biomedical issues from a Christian theological and ethical perspective. He has also made significant contributions to the radical human enhancement debate. In an article titled "Christian Transhumanism,"[3] he also points to Dante's use of the term.

[1] Julian Huxley, *New Bottles for New Wine* (London: Chatto & Windus, 1957), 17.

[2] Vita-More's careful etymological work, done in the late 1980s to the mid-1990s, is published in numerous places, e.g., Vita-More, "Life Expansion: Toward an Artistic Design-Based Theory of the Transhuman/Posthuman," PhD diss., (University of Plymouth, 2012), 78–79. http://hdl.handle.net/10026.1/1182. See also the 2008 version of Natasha Vita-More, "The Transhumanist Manifesto," especially footnote 2, which succinctly summarizes her findings. https://natashavita-more.com/transhumanist-manifesto/.

[3] "Christian Transhumanism," in *Religion and Human Enhancement: Death, Values, and Morality*, eds. Tracy Trothen and Calvin Mercer, in *Palgrave Studies in the Future of Humanity and*

C. Mercer, T. J. Trothen, *Religion and the Technological Future*,
https://doi.org/10.1007/978-3-030-62359-3_3

Trasumanar was Dante's word to describe the glorious transformation that Christians experience as they ascend into God's presence. Cole-Turner paraphrases Dante: "To go beyond the human is something that cannot be described in words."[4] Cole-Turner reclaims the Christian roots of "transhumanism," which he argues go all the way back to biblical themes. Here is just one biblical example: "What we will be has not yet been revealed. What we do know is this: when he (Christ) is revealed, we will be like him."[5] The debate about the origins of the terms reflects broader and deeper questions about whether the nature and mission of the religions position them to embrace radical human enhancement, questions that will concern us in the upcoming chapters.

Pushing forward from the origins of words and concepts, transhumanism as a movement began with the extropian movement in the late 1980s through the work of philosopher Max More[6] and others.[7] The term "extropy" was chosen by More to reflect a desire for improvement and growth, in contrast to entropy, which measures disorder within a system. Today, transhumanism is generally understood as an intellectual and cultural movement that advocates the use of a wide range of increasingly powerful technologies to radically enhance human beings. Transhumanists look for the convergence of these technologies to make it possible to control human evolution by using technology to enhance human physical, mental, affective, moral, and spiritual abilities and to ameliorate aspects of the human condition regarded as undesirable, such as disease and aging.[8]

Organizations advocating for development and use of these varied and fast-developing technologies are relatively small, but powerful, and getting

Its Successors, series eds. Calvin Mercer and Steve Fuller (New York: Palgrave Macmillan, 2017; paperback, 2019).

[4] Paraphrase of Dante, *Paradiso*, canto 1, line 70, in Cole-Turner, "Christian Transhumanism," 37.

[5] In the New Testament, 1 John 3:2.

[6] At that time Max More's name was Max O'Connor. He changed his name to reflect his desire for the extropian commitment to "more" life, freedom, and intelligence.

[7] The history is well documented and analyzed by James Michael MacFarlane, *Transhumanism as a New Social Movement: The Techno-Centred Imagination*, in *Palgrave Studies in the Future of Humanity and its Successors*, series eds. Calvin Mercer and Steve Fuller (New York: Palgrave Macmillan, 2020). On the history, see especially 25–54.

[8] More provides a good introduction to the philosophy of transhumanism, its various precursors, its relationship to humanism, some contemporary expressions, and misconceptions in "The Philosophy of Transhumanism," in *The Transhumanist Reader: Classical and Contemporary Essays on the Science, Technology, and Philosophy of the Human Future*, 3–17, eds. More, Max and Natasha Vita-More. (West Sussex, UK: Wiley-Blackwell, 2013). In a brief section, page 8, he contends that religion and transhumanism are not necessarily incompatible. *The Transhumanist Reader: Classical and Contemporary Essays on the Science, Technology, and Philosophy of the Human Future* is an excellent collection of transhumanist writings, including, as the subtitle indicates, some classic pieces by, e.g., Robert A. Freitas, Jr. (nanotechnology), Ralph Merkle (nanotechnology), Marvin Minsky (artificial intelligence), and Hans Moravec (robotics). A recent collection of articles by scholars and others on the history, philosophy, religion, and technology of transhumanism is edited by the chair of the California Transhumanist Party, Newton Lee, and titled *The Transhumanism Handbook* (New York: Springer, 2019).

stronger; together they help form the transhumanist movement. Although changing, transhumanism has been largely led by white men from Europe and America, a fact addressed later in our ethical reflections. The leading secular transhumanist organization is "humanity+,"[9] with the website headline, "Don't limit your challenges, challenge your limits." Although humanity+ has only a few thousand formal members, transhumanist ideas, and especially the technologies associated with them, are increasingly working their way into the public's awareness. Although now only a tiny blip on the political radar, there is in the United States a small political party structure that could provide the context for intense debates about policy and funding in the coming decades.[10] Later in this chapter, we provide detailed information about two other transhumanist organizations that are religious in focus.

In one sense of the word, a transhuman is still a human, but one whose capabilities go beyond those of "normal" humans. One could argue we are already transhuman, since we interface so heavily with technology. Radically enhancing ourselves into transhumans may not be the end of the matter. At some point, the enhancements may be extreme enough that the human may evolve into the posthuman. *Homo sapien* may become *techno sapien* or *Homo Deus.* The transhuman will turn out to be an intermediary step between "normal" humans and beings that are not human, i.e., are posthuman. We entertain the possibility of the posthuman in later chapters on mind uploading and superintelligence.

Narrowly conceived, the term transhuman refers to that which is still human, albeit greatly enhanced. However, the term is sometimes used more loosely, in the broader sense of (1) sentient beings that develop from humans but to such a degree that they are no longer human in any real sense, and (2) sentient beings that develop apart from humans. It is in this broader sense that we use the term transhumanism. So, in the coming chapters we discuss, for example, the development of advanced artificially intelligent robots, which develop apart from any direct evolution of *Homo sapien.*

We end this section with a suggestion to read this clever piece, "A Letter to Mother Nature: Amendments to the Human Constitution,"[11] from Max More, the influential transhumanist. This letter, and its important amendments, captures well the sentiment of many transhumanists. We will encounter More again in the later chapter on cryonics. Here are some key excerpts from this letter to Mother Nature. Notice how transhumanism tends toward a view of human nature as malleable, as a work-in-progress.

> Mother Nature, truly we are grateful for what you have made us. No doubt you did the best you could. However, with all due respect, we must say that you have

[9] http://humanityplus.org/.

[10] "U.S. Transhumanist Party—Official Website." http://transhumanist-party.org/. For a discussion of how it fits into the history and current presence of the transhumanist movement, see MacFarlane, *Transhumanism as a New Social Movement*, 155–162.

[11] http://strategicphilosophy.blogspot.com/2009/05/its-about-ten-years-since-i-wrote.html.

> in many ways done a poor job with the human constitution. You have made us vulnerable to disease and damage. You compel us to age and die—just as we're beginning to attain wisdom. You were miserly in the extent to which you gave us awareness of our somatic, cognitive, and emotional processes. You held out on us by giving the sharpest senses to other animals. You made us functional only under narrow environmental conditions. You gave us limited memory, poor impulse control, and tribalistic, xenophobic urges. And, you forgot to give us the operating manual for ourselves! ... What you have made us is glorious, yet deeply flawed. ... We have decided that it is time to amend the human constitution. ... We do not do this lightly, carelessly, or disrespectfully, but cautiously, intelligently, and in pursuit of excellence. We intend to make you proud of us. Over the coming decades we will pursue a series of changes to our own constitution, initiated with the tools of biotechnology guided by critical and creative thinking.

Next, there follows seven amendments to the human constitution. More's letter ends in the following way:

> These amendments to our constitution will move us from a human to a transhuman condition as individuals. We believe that individual transhumanizing will also allow us to form relationships, cultures, and polities of unprecedented innovation, richness, freedom, and responsibility. We reserve the right to make further amendments collectively and individually. Rather than seeking a state of final perfection, we will continue to pursue new forms of excellence according to our own values, and as technology allows.

The letter is signed "Your ambitious human offspring."

Monotheistic Religions and Karmic Religions

Scholars specializing in the monotheistic religions of Judaism and Christianity have been most active in evaluating technological enhancement of humans. However, we want this textbook to play a role in opening up the conversation beyond these two traditions.[12] To that end, we here paint in very broad and introductory strokes some key theological concepts in all religions to which we refer as we reflect on radical human enhancement in the coming chapters. Our

[12] Scholars of Buddhism, Daoism, Hinduism, Islam, and Jainism contributed chapters to *Religion and the Implications of Radical Life Extension,* eds. Derek Maher and Calvin Mercer (New York: Palgrave Macmillan, 2009; paperback, 2014); and *Transhumanism and the Body: The World Religions Speak*, eds. Calvin Mercer and Derek Maher, in *Palgrave Studies in the Future of Humanity and Its Successors,* series eds. Calvin Mercer and Steve Fuller (New York: Palgrave Macmillan, 2014). We expect the karmic religions will become increasingly engaged in the conversation, and publications like the following suggest this may happen sooner rather than later: Robert M. Geraci, *Temples of Modernity: Nationalism, Hinduism, and Transhumanism in South Indian Science* (Lanham, Maryland: Lexington Books, 2018); and Zhange Ni, "Reimagining Daoist Alchemy, Decolonizing Transhumanism: The Fantasy of Immortality Cultivation in Twenty-First Century China," *Zygon: Journal of Religion and Science* 55, no. 3 (August 2020): 748–71. https://doi.org/10.1111/zygo.12634.

hope is that our very preliminary reflections will prompt increased discussion from all the religions about human enhancement. We underscore that our goal in the textbook is not to give comprehensive summaries of the religions, but, rather, to identify a few important themes that have been or could be useful in thinking about enhancement technologies. These themes are meant to be illustrative, not exhaustive.

Terminology categorizing the world's religions is problematic. "Eastern religions" and "Western religions" are sometimes used, but these terms are quite broad and emphasize geographical origin, which is no longer as relevant with globalization spreading the religions across the world. We chose the terms monotheistic religions and karmic religions, recognizing that these terms are also imperfect and do not capture all the ways of being spiritual or religious.

Monotheistic religions include the "Abrahamic faiths," Judaism, Christianity, and Islam. We use the term "karmic religions" to refer to Hinduism, Buddhism, Daoism, Jainism, and Sikhism, although each religion understands karma and rebirth (to be discussed in more detail later) in varying ways.[13] We draw ideas especially from India and the religions of Hinduism and Buddhism. We will also refer to perspectives derived from Indigenous Spiritualities and more secularized spiritualities.[14]

There are many hybrids of all these religions, and we are aware we have not addressed all of the religions or ways of being religious. The generalizations we do make, for example, about a particular religion are just that, generalizations that are helpful in reflecting on radical human enhancement but do not apply to every strand or expression of the religion. We encourage you to further explore any religion or way of being religious or spiritual that captures your interest. Information provided about religions in this textbook is meant as a beginning study only.

This textbook presumes no background knowledge about religion; everything about the faith traditions needed for the discussion in the following chapters is provided. Excellent resources are available for those desiring a more thorough treatment of the world's religions than provided in this textbook.[15]

Conceptualizations of the Divine

Throughout history and across the planet, human beings have conceptualized the divine in many ways. "Theism" is from the Greek word *theos* (god) or *theoi*

[13] Other religions include those originating in East Asia, including Confucianism and Shinto. Like the monotheistic religions, Zoroastrianism originated in the Middle East but is not strictly monotheistic.

[14] Indigenous traditions, sometimes called tribal or basic religions, include African, Native American, Innuit, and others.

[15] The most detailed introduction to the world's religions is probably David S. Noss and Blake R. Grangaard, *A History of the World's Religions*, 14th ed. (Abingdon, UK: Taylor and Francis, 2017). An excellent shorter introduction is Michael Molloy, *Experiencing the World's Religions: Tradition, Challenge, and Change*, 8th. ed. (New York: McGraw-Hill, 2020).

(gods). The two theisms that will concern us most are pantheism (*pan* is Greek for "all" or "everything") and monotheism (*monos* is Greek for "one" or "single").

While not all strands of karmic religions are pantheistic, this is an important perspective in those traditions. Pantheism is the belief that God is equivalent to all of reality. God is, by definition, everything, and everything is God. Pantheism can be difficult to understand for many who do not follow a karmic religion. Pantheism is not polytheism (*poly* is Greek for "many"), which is a belief in many gods, particularly characteristic of ancient religion, as in the ancient Greek and Roman pantheon of deities and often associated with aspects of nature. With pantheism, God is not any particular part of nature or reality. God is everything in total and everything in total is God.

Looked at from another angle, pantheism is the idea that there is only one "thing." In other words, everything is really the same thing. In the deepest sense of reality, there are no distinctions. There is only one reality, and it is all the same. That one "thing" is, of course, called God in this pantheistic model. When ignorance is vanished, and truth fully expressed, there is only God. Those not immersed in Asian cultures, where pantheism is an underlying mindset, may react this way: "Pantheism is an interesting idea, but it is obviously incorrect. This chair is clearly distinct from that table, and both are clearly distinct from my body." Not so fast.

Enter the concept of *maya*, Sanskrit for "illusion." According to this perspective, we are mistaken to think reality is made up of different things. We are under the power of a pervasive illusion. The gurus of this model of the divine liken it to a dream. When we dream and do not know we are dreaming, if it is a nightmare, our heart beats fast and our blood pressure rises. We are convinced the nightmare is real. Then, we arrive at that interesting moment when we transition to the awake state, realizing, "Oh, it was just a dream."

The karmic religion sages working from a pantheistic model contend that we are dreaming a world of distinctions. It is powerful, sophisticated, detailed, and very convincing, but it is not real, at least it is not ultimately real. The world of distinctions is an illusion. In pantheism, even the many Gods in Hinduism are manifestations of the oneness. There is only one. There is only God. God is everything, and everything is God.

Every religion identifies what it considers to be the basic human predicament. For Hinduism, to give one example, the problem is ignorance. Salvation, then, comes with spiritual knowledge that brings enlightenment, waking up from the illusory dream and coming fully into the knowledge, the realization, of oneness with God. This pantheistic understanding of the divine has implications for how human beings and their salvation are conceived. We will address that in a moment, but first it will be helpful to contrast pantheism with monotheism, the dominant model of the monotheistic religions: Judaism, Christianity, and Islam.

Monotheism posits one deity, a God who creates all of reality and is distinct from that reality, including human beings. Although there are plenty of variations, the attributes of the one God are usually understood to include omnipotence (all-powerful), omniscience (all-knowing), and omnibenevolence or omnibeneficience (all-good). Some followers see God as both immanent (existing within humans, animals and/or nature) and transcendent (being apart from and greater than), whereas others see God as transcendent only. The human predicament is that the human creatures are estranged from God through sin or acting contrary to divine intention. Salvation entails reconciliation through various means, such as divine grace, confession, atonement, and restitution.

Theological Anthropology

"Theological anthropology" is a fancy word for a religion's doctrine, or belief, about human beings. Who are human beings, what is their relationship to God, and how can they journey to salvation?

Karma and Reincarnation

An important tradition in Hinduism, to pick one karmic religion example, asserts that human beings, caught up in the illusory world of distinctions, have a soul (Sanskrit *atman*). The soul is the infinite center of every life, distinct from the body and a reservoir of being that is unrestricted in consciousness and bliss. When salvation is achieved, the ultimate truth is revealed that the soul and God are really one and the same, because in pantheism God is everything and everything is God.

The doctrines of karma and reincarnation explain how the human soul arrives at saving, enlightenment knowledge. *Samsara* is Sanskrit for "wandering" or "going through." Commonly known as "reincarnation," but also called rebirth or transmigration of the soul, the doctrine teaches that in the world of illusion the human soul transmigrates through successive lives until it achieves liberation from the cycle. So, there is an aspect of every being—the soul, mind, or consciousness—that is eternal and indestructible and is reborn as human, animal, and heavenly or hellish beings until final liberation (Sanskrit, *moksha*) from the "wheel of rebirth."

Poetically put by the book perhaps most beloved in India, the *Bhagavad Gita*, "Worn out garments are shed by the body; worn out bodies are shed by the dweller."[16] This notion of the soul's rebirth through successive lives is most famous in India, although it was also held in the ancient Near East and other cultures. Our purpose here is not to engage the distinctions between, for example, the varying notions of reincarnation in Hinduism and Buddhism. Rather, with regard to assessing radical human enhancement, we will reflect more broadly on how this central belief of reincarnation might push in a different direction than the monotheistic religion view of one life and one afterlife.

[16] *Bhagavad Gita* 2:22. Edwin Arnold, trans., *The Bhagavad Gita* (Kansas City: Scholar's Choice, 2015).

Integral to the doctrine of reincarnation is the law of karma. The Sanskrit word literally means "action" or "deed," and it refers to the idea that one's actions (or inactions) influence, even determine, the status of the lives into which a soul is born. Science has taught us the law of cause and effect. Every event in the physical universe has a cause, and every cause has determinate effects. The law of karma extends cause and effect to the moral and spiritual realms. Moral, spiritually good actions in this life promote rebirth into a next life that is closer to liberation.

While we are not fully teasing out the differences between Hinduism and Buddhism on these matters, we should note that Buddhism has its own special interpretations of the traditional doctrines of karma and reincarnation. For Buddhists, reincarnation occurs without any actual soul-substance passing from one life to the next.

Monotheistic Views of Soul, Body, and Physicality

In the monotheistic religions, there is disagreement about what constitutes a human being and the role soul has in that constitution. Judaism and Christianity, for example, in the main depict a human being as a psychosomatic unity of body and soul. Soul is not a distinct and separate "part" of what constitutes a person. However, influenced by ancient Greek and other dualistic traditions, one Christian view, popular in many conservative Protestant circles, asserts that human beings do possess a separate soul, which is the focus of salvation.

The predominate view, however, that of a psychosomatic unity of body and soul, leads to the doctrine of resurrection, which entails the raising of an integrated being, not just a soul. In other words, there is a clear difference between immortality of the soul only and resurrection of the unified person, which includes the (transformed) body.[17] An immortal soul that transmigrates from one bodily incarnation to another is a notion prominent in Hinduism (as we discussed earlier), neo-Platonic groups, and ancient Gnosticism.

The belief that the body and soul can be separated is a dualistic belief. We need not complicate matters in making this point, but the influential seventeenth century philosopher, René Descartes, also distinguished mind from body in a way that supports dualistic thinking. An immortal soul is a belief central to many religious traditions. Our point is that a dualistic view of the person is not the view in the ancient Jewish and Christian scriptures and has not been the anthropology affirmed in mainstream Christian theology. Today, many Protestant Christians, especially those with a conservative bent, use the biblical language of resurrection, but their ideas about the afterlife are much more grounded, albeit unknowingly, in a dualistic framework derived from sources outside the Bible.

[17] For a valuable collection of articles on resurrection and science, see Ted Peters, Robert John Russell, and Michael Welker, eds., *Resurrection: Theological and Scientific Assessments* (Grand Rapids: William B. Eerdmans, 2002).

The notion of the psychosomatic unity of body and soul, with the attendant notion of resurrection, is underscored by the importance of physicality. Physicality is so central to this vision that it merits putting it in a larger biblical and theological context. The Christian religion illustrates the emphasis on physicality. The incarnation of God in Jesus Christ, the full uniting of the divine and human, is one of the central doctrinal tenets of Christianity. God became flesh; God embraced physicality. Raw material for this central doctrine is found in the first chapter of the Christian Gospel of John. The eternal, divine *logos* (Word) is fleshed out in Jesus.[18]

In the first few centuries of the Christian church, this idea was hammered out in creeds and councils. Christianity rejected the docetic (from the Greek word meaning "to seem") heresy that taught that Jesus only "seemed" like a human being, but really was not. Finally, in the Council of Chalcedon (451 CE), the church formulated the position that has survived the centuries, despite occasional dissent. Jesus is fully divine and fully human, and his humanity includes, of course, embodiment.

The importance of the body even shows up in one of Christianity's central rituals, the Eucharist, in some traditions called the Lord's Supper or Holy Communion. In this ritual, Christians eat bread and drink wine.[19] A variety of interpretations of this ritual can be found among Roman Catholics and Protestants, and even among various Protestant denominations. In all of them the bread, whether understood literally or symbolically, is the body of Christ. Indeed, the Christian church is called the "body of Christ," again emphasizing the importance of the corporeal.

The embrace of physicality, profoundly exhibited in the Christian doctrine of the Incarnation, is also reflected in Judaism. In the first chapter of the first book of the Bible, God creates the entire physical world and affirms it as "very good."[20] This affirmation carries through to and is part of the eschatological vision of a "new heaven and a new earth."[21]

Christianity teaches the possibility of an afterlife in which one is immortal, living with God for eternity, although in a transformed state. What this transformed state looks like is debated among followers of this religion with many claiming that the afterlife is a divine mystery that is to be known only after death. Some believe that one's destiny after death is determined by the way in which one lived. Others see salvation as less dependent upon works and more dependent upon God's grace. The important thing for matters discussed in this textbook is the general belief in the possibility of an eternal life that will be more glorious than earthly life and will entail, in the dominant Christian view, embodiment.

[18] John 1:1, 14.

[19] In Christian denominations that forbid alcohol, or see alcohol as excluding some members, grape juice is usually substituted for wine.

[20] Genesis 1:31.

[21] Isaiah 65:17; Revelation 21:1.

"Theosis" or Deification

Whether the human being is understood as "having" a soul or "being" a soul, the monotheistic traditions generally rebel against identifying God with the human or with anything else that God creates. There is only one God, and it is idolatrous to make any part of the created order divine.

Having said that, we need to consider one very important exception in the monotheistic tradition of Christianity. Gaining a good bit of attention of late is an important theological notion in Eastern Orthodox Christianity, one of the three main branches of the Christian religion.[22] The Eastern Orthodox Christian doctrine is called *theosis* and is well expressed in the often-quoted words of Athanasius of Alexandria (296/298–373 CE), "He [Jesus Christ] was made man that we might be made god."[23] The idea is also affirmed by Mormonism, a Christian movement discussed later in this chapter.

Theosis, sometimes called deification, refers to a view of salvation whereby one is transformed, or divinized, and elevated in some significant way into the life of God. The discussions about (and terminology of) *theosis* are quite technical and complex, perhaps in part due to the effort to stop short of attributing full divinity to a human, which would be heresy in a monotheistic religion.

Theosis is interesting in that it is perhaps the best example of a theological concept in an Abrahamic religion that compares closely to pantheism in karmic religions. There is an important difference, however. Traditionally, at least in Hinduism, a soul can move through thousands or more rebirths before arriving at liberation. Buddhism, in a reformation of Hinduism, asserted that it is possible for anyone to become enlightened in this lifetime, although it can take many reincarnations. *Theosis*, however, is a spiritual process that theoretically occurs in a single lifetime as the Eastern Orthodox Christian pilgrim moves through three stages: (1) purgative or purification, (2) illumination, and, finally, (3) *theosis*, deification or unity with God.

Created Co-Creators[24]

Christian theologian Philip Hefner proposed the interesting notion that human beings are created co-creators with God.[25] Human beings, created in the image of God (Latin, *imago Dei*), are charged in the first chapter of the Bible with being stewards of the created order. Human beings are responsible for tending the garden, keeping it beautiful, and flourishing. So, using God-given talents, people work with God as created co-creators in the creative process to make anew.

[22] The other two branches in this religion are Roman Catholic and Protestant.

[23] Athanasius of Alexandria, *On the Incarnation of the Word*, trans. John Behr, *Popular Patristics* Series 44 (New York: St. Vladimir's Seminary Press, 2011), section 54.3, p. 167.

[24] For a discussion of the created co-creator principle and Christian theology, see Stephen Garner, "Christian Theology and Transhumanism: The 'Created Co-creator and Bioethical Principles," in *Religion and Transhumanism: The Unknown Future of Human Enhancement*, eds. Calvin Mercer and Tracy Trothen, 229–43 (Westport, CT: Praeger, 2015).

[25] *The Human Factor: Evolution, Culture, and Religion* (Minneapolis, MN: Fortress, 1993), 27.

Hefner's theological reasoning goes like this. If we are created in the image of God (Latin *imago Dei*), as expressed in a Jewish creation story,[26] and if one of the divine qualities is creativity, then human beings are meant to create so long as this creating is done with humility (knowing they have a propensity to make mistakes and to sin), the intent to do good (repair the world), and with the communal and ongoing discernment of God's will, as best that will can be discerned. In other words, we are meant to do good and to care for creation, ourselves, and our neighbor. Professor Trothen explains:

> Humans, as made in the likeness of God, have been given creativity to be used in divine service. Theologian Philip Hefner's proposal of humans as created co-creators helps us to complicate the caution not to "play God" in the realm of science and technology. While we ought to be prudent and aware of our abilities to mess up, we also have extraordinary capacity to improve life with God's help. The *imago Dei* suggests a divine mandate to create for the good. This is a risky venture requiring humility and some audacity. Theological ethicist Grace D. Cumming Long develops the ethical principle of creativity making a strong case for the necessity of creativity if we are to reimagine and recreate the world as just and compassionate .... We have means to assess the damage we have done and continue to do to the environment, and there are possible correctives if we collectively have the political will.[27]

The concept of created co-creators lays the theological ground for asserting that God can and does work through and with people, as humble and courageous partners, in developing technologies, perhaps very powerful ones, for good. To say it a different way, technology can be a means of grace, just like the hands of the traditional doctor. The concept is similar to a Jewish notion, *Tikkun olam*, translated "repair the world." A justice oriented "co-evolution" could be considered the secular version.[28]

To put it yet another way, the material world can be a domain of God's graceful, creative action. Lurking behind this theology is, of course, an optimistic view of how technology can be both potent and good in the hands of God and her/his created co-creators. So, the fundamental assessment of nature's potential for good, in the monotheistic religions, is a deeply biblical concept, going back to the very important creation stories in Genesis. Following the sixth day of creation, God declared that everything is "very good."[29] There is no hint of evil, suffering, or fallenness in the creation story in the first two chapters of the Hebrew Bible. Everything is very good, including human physical, cognitive, affective, moral, and spiritual abilities.

[26] Genesis 1:26.

[27] Tracy J. Trothen, "Christian Anthropology: Doing Good Through Science and Technology," in *Technology and the Image of God: A Canadian Conversation* (Canadian Council of Churches, Faith and Life Sciences, 2017), 18–19.

[28] "Co-evolution with technologies" is used by MacFarlane, *Transhumanism as a New Social Movement,* 2.

[29] Genesis 1:31.

If the Bible had ended at Genesis, Chap. 2, the story would be of a good God creating an incredibly wonderful world with no hint of evil and suffering. Human beings are able and good-faith stewards of that world. Wonderfully situated in the lovely Garden of Eden, the human creatures are charged with naming the animals and tending the garden. When God comes walking through the garden in the cool of the day, Eve and Adam, the first humans created, run out to greet their creator. Then, we arrive at Genesis, Chap. 3, where the humans freely, deliberately, and painfully make irresponsible decisions that bring heartache and pain. Adam blames Eve, Eve blames God, Cain kills Abel, they build the Babel tower, and that age's version of nuclear holocaust comes in the form of a worldwide flood.

Most assuredly, the above interpretation of the biblical stories of creation as originally good and human beings as created co-creators with God is not the only interpretation. Earlier, we referenced how some Christian thinking was influenced not by this Jewish background, but by Greek-inspired dualistic philosophies that postulated a soul separate from a physical body. That same dualistic thinking was sometimes extended to judge physicality as bad or evil. Such a dualistic view could very well lead to the conclusion that the physical world is evil by nature and God's work is to save the faithful from that evil world. In such a worldview, technology is part of the evil world and is not seen as a means of God's action in the world.

The Spectrum of Theology

While labels are often libels, doing more harm than good, sometimes labels and broad generalizations can be useful in beginning to understand ideologically based movements or trends. Used cautiously, terms like "liberal" and "conservative" can be useful in framing judicial philosophy, political leaning, and economic theory. Theological commitment in religion also often falls along a liberal-conservative continuum.

"Liberal" refers to the left side of the theological spectrum, a side some prefer to call "progressive." "Conservative" refers to the right side of the spectrum. Conservative, as we are using it, includes fundamentalism, which would be the extreme right. The next page contains a helpful table showing theological generalizations of the liberal-conservative continuum.

As noted in the continuum (Table 3.1), theological liberalism contends that God affirms and works through the natural world to foster peace and justice, to bring God's Kingdom/Kin-dom ever closer to fruition in this world. This contention generally puts liberals in an open—even if careful—stance toward science as providing knowledge about God's creation and enabling technology that benefits humanity.

So, the progressive wings of the religions, with their generally favorable attitude toward science, are positioned to adapt religion to radical enhancement. For example, a recent book about Islam and transhumanism[30] illustrates

[30] Roy Jackson, *Muslim and Supermuslim: The Quest for the Perfect Being and Beyond*, in *Palgrave Studies in the Future of Humanity and Its Successors*, series eds. Calvin Mercer and Steve Fuller (New York: Palgrave Macmillan, 2020).

Table 3.1 Theological Continuum

	Conservative	*Liberal*
The Divine	Theocentric (God-centered), low view of human beings as sinful and weak	Anthropocentric (human-centered), high value on human ability given by God, optimistic about human capability, God works through the created order
The World	Otherworldly, emphasizing the reality and importance of the realm above and beyond the natural, such as heaven, hell, souls, and the afterlife	This-worldly, emphasizing the here and now, more occupied with immediate, present social and other problems, rather than with future spiritual destiny
Revelation	Religious truth comes from God through special communication, such as sacred writings (fundamentalist), institutions and traditions (Roman Catholic), or personal religious experience (Pentecostal/ charismatic)	Traditions and texts are important, but truth in them is appropriated and discerned by the God-given rational ability of human beings
Tradition	Traditionalists, valuing religious understanding handed down from previous generations, generally opposed to change	Revisionist, sees the necessity for revising and updating traditional notions in light of changing circumstances and new knowledge, reluctant to claim infallibility for any doctrine
Attitude	Dogmatic, committed to certain indisputable beliefs that are not open for questioning, modification, or debate, because they are supernaturally revealed	Pragmatic, interested in what works, what solves human problems and meets human needs

Adapted from Calvin Mercer's *Slaves to Faith: A Therapist Looks Inside the Fundamentalist Mind* (Westport, CT: Praeger, 2009), 47, and based on George C. Bedell, Leo Sandon, Jr., and Charles T. Wellborn, *Religion in America* (New York: Macmillan, 1975).

how a "creative and explorative" approach to the religion positions Islam as an inspirational and productive interlocutor in the debate about radical human enhancement.[31] If a positive stance toward science was the only consideration, we would conclude that religious liberals are supportive of the transhumanist agenda, since it is science-based and with the mission of improving the human condition through technology.

However, liberals are also concerned about social justice. They worry that enhancement therapies and technologies are going to be the privilege of the wealthy and powerful. Also concerning to liberals is that this privileged class may dictate the nature and purposes of these technologies. Consequently, many liberals are extremely cautious about, and even opposed to, enhancement technology. We explore these issues in more detail in the following chapter that addresses ethics.

[31] Ibid., 5.

Liberal and Conservative Reactions

Many religious conservatives, and especially fundamentalists, tend to see the world as the domain of temptation and evil. God's redemptive acts that lead us to peace and justice are in spite of the sinful world, rather than through the leaders and systems of the world. Science and technology are, at best, enterprises of sinful humans with some positive breakthroughs due to God's grace and, at worst, the work of the devil.

Liberal optimism has suffered a huge setback with the two world wars, one of which included a horrendous Holocaust, and other global atrocities. Now, with a pandemic, terrorism, wide-scale racial justice protests, and other global violence a part of our recent past, where is the evidence that God is working through human beings and through the world? It seems we are not always moving toward increased community, peace, and justice.

With the world seemingly bent on evil and destruction, there arose on the right side of the theological spectrum a Christian Protestant movement known as "neo-orthodoxy." Championed by the influential German theologian Karl Barth and American theologian Reinhold Niebuhr, neo-orthodoxy began as a critique of the left emphasis on human goodness and preached our need to depend on God's grace and God's radical intervention to save sinful humanity. Neo-orthodoxy was not right-wing fundamentalism, and perhaps is best interpreted as a New Reformation Theology that was a mix of liberal and conservative and something more. While religions tend to develop liberal and conservative poles, most movements are somewhere along the continuum between the left and right extremes.

Another theological response to the growing social disquiet of the post-World War II decades came from the more radical left in the form of liberation theologies, with third world, feminist, and Black being prominent examples. These leaders include Gustavo Gutiérrez, Rosemary Radford Ruether, and James Cone. These theologies grew out of marginalized voices that championed social justice and strongly critiqued the status quo, including the theological status quo. Liberational theologians looked anew at the Bible, with its themes of exodus from oppression and slavery, courageous prophetic preaching for social justice, and privileging of the poor and dispossessed in Jesus' ministry.

Process theology is another example of a progressive way of formulating doctrine that can be employed in a positive assessment of radical human enhancement. This approach is based on the work of English mathematician and philosopher Alfred North Whitehead. At a basic level, process thinking begins with the real world, wherein everything changes all the time. Trees grow, buildings crumble, the sun rises, teeth decay, relationships change, and so it goes for everything, all the time. Christian theologians, such as John B. Cobb and Charles Hartshorne, influenced by Whitehead, argued that since everything changes, God does as well, an idea we explore in a later chapter.

The idea that change is intrinsic to the nature of reality has wide support. In Western philosophy, its most well-known ancient Greek advocate is Heraclitus (c. 535–475 BCE) who famously said, "You cannot step twice into the same stream."[32] Buddhism reinforces this point about change. One of the very important Buddhist "Three Signs of Being"[33] is the law of impermanence (Pali, *anicca*, literally "nonpermanent"). While impermanence is central in Buddhism, it plays an important role in some schools of thought in Jainism and Hinduism, where it is found in important Hindu scriptures, such as the *Rigveda*, *Upanishads*, and *Bhagavad Gita*.

A key underlying question for the religions is how flexible they are in the face of modern world developments, such as in the science and technology of radical human longevity. Conservative attachment to tradition and dogmatic attitude resists evolving religion in ways that accommodate changing circumstances, while liberalism works hard to find ways to adjust its theology. Liberal religion is nimble; it flexes and adapts to new ideas and ways, although sometimes at the expense of core beliefs, according to many conservative critics.

Often implicit in the liberal embrace of transhumanist programs is a positive assessment of technology as a means of divine action in the world. The moral status of technology is a subject of considerable academic debate.[34] The "instrumental view" of technology is the idea that technology is value-neutral and can be utilized to morally positive or morally negative ends, depending on the intentionality and motive of the human guide. However, several well-regarded philosophers have shown that technology is imbued with the values of utility and efficiency. More will be said about values and technology in the next chapter on ethics.

Drilling down below these various theological positions, we reach fundamental questions. Are we basically optimistic or pessimistic about the human spirit, about human intentionality? Relatedly, do we think, to put it bluntly, that we are going to survive in good fashion or do ourselves in with these new technologies? Splitting the atom led to electricity for our homes and also created a means for the end of civilization and planetary collapse. While such questions can be asked by any human being, these questions are religious questions. Put theologically, does a religion's theological anthropology understand human nature as good or bad or somewhere in-between? Liberals tend to drop down on the side of the goodness and potential of the human creature.

[32] Plato, *Cratylus* 402a, in *Plato in Twelve Volumes*, vol. 12, trans. Harold N. Fowler (Cambridge, MA, Harvard University,1921). http://www.perseus.tufts.edu/hopper/text?doc=Perseus%3Atext%3A1999.01.0172%3Atext%3DCrat.%3Asection%3D402a.

[33] The other two are suffering and no-self.

[34] For a review of options, see Maarten Franssen, GertJan Lokhorst, and Ibo van de Poel, "Philosophy of Technology," *Stanford Encyclopedia of Philosophy*, last updated September 6, 2018. https://plato.stanford.edu/entries/technology/; Thomas A. C. Reydon, "Philosophy of Technology," *Internet Encyclopedia of Philosophy*, accessed November 2019. https://www.iep.utm.edu/technolo/; and James Michael MacFarlane, *Transhumanism as a New Social Movement*, 5–8.

Conservatives tend to emphasize our fallen nature and are, consequently, cautious about putting too much control, including technology, into our hands.

Religious Transhumanism

Transhumanism as a movement was initially populated primarily by secularists. As it gains momentum and the enhancement technologies become widely known, if not utilized, transhumanism will continue to include more people of faith from all the religions.[35] It is important to be clear about what it means to be a religious transhumanist. Most transhumanists see the body as a machine that needs repair and improvement. That is a secular, mechanistic view of who we are. Although it may be slowly changing in light of a more holistic model, much modern medicine is similar to transhumanism in its view of the body as made up of mechanistic parts that can be fixed.

Most religions reject this simple reduction of who we are. Physicality is not all there is; our embodiment is one pervasive aspect of our being, but not the only one. Many karmic traditions attend to the soul and its cosmic welfare. Even Buddhism, with its "no soul" view, understands that reincarnation and the law of karma place our lives in a larger cosmic context than that afforded by a strict materialist worldview. Indigenous spirituality understands all life as profoundly intraconnected and interconnected. Healing cannot occur without attending to the whole person.

In the monotheistic religions, God as creator breathed life into the human creatures. The Bible does not speak about our bodies in isolation from ourselves as persons animated by God. The worldview of materialism is rejected by Judaism, Christianity, and Islam that stem from the same biblical tradition. So, if there is a religious transhumanism that could sanctioned by the Abrahamic religion followers, then the secular, materialistic program must be somehow incorporated into the religious vision.

Because the religions understand physicality as an integrated aspect of who we are, and care about our physical well-being, the world faiths are positioned to embrace enhancement to some degree should they choose. Professor Trothen summarizes this point:

> Religion places a high value on healing and minimizing suffering. For example, Christians are guided by the many stories of Jesus' healings. Some unavoidable creaturely suffering is seen as valuable, potentially contributing to spiritual growth and community. Buddhists are committed to reducing suffering by loosening one's attachment to the self as an identity, in light of the Buddhist notion of "no-self." Hindus see deserved suffering as necessary to ward off bad karma but undeserved suffering can be minimized. Daoism promotes a variety of practices for

[35] MacFarlane, *Transhumanism as a New Social Movement,* 177–201 discusses various aspects of transhumanism and religion.

> health and long life. And, in Judaism the preservation of life is the highest mitzvah.[36]

Ron Cole-Turner, a Christian theologian introduced earlier in this chapter, argues that, indeed, Christian transhumanism differs significantly from secular notions of transhumanism, especially in its emphasis on the need to let go of self (an idea that resonates with Buddhists) in order to find one's true self in relation with God. With this understanding, transhumanism is properly a Christian concept, he argues. In his article entitled "Christian Transhumanism," he writes:

> Christian transhumanism is not an accommodation to our age. It is instead an affirmation of the radically transformative nature of the hope that lies at the heart of a Christian view of humanity and the cosmos.[37]

In a statement that is typical of religious transhumanists of all religions, Cole-Turner is quick to point out that a larger self-sacrificial vision should motivate people of faith, even if they identify as transhumanists: "The pathway to life is not found in preservation and extension but, paradoxically, in 'self-denial' and in willingness to lose one's own life."[38]

Mormon Transhumanist Association

We have made the case that the religions are diverse and flexible enough to have elements that can embrace radical human enhancement. Indeed, we find that flexibility exampled by two Christian transhumanist organizations. Organizationally, the oldest pro-transhumanism religious group is the Mormon Transhumanist Association (MTA),[39] a Christian movement started in 2006. "Transfigurism" is the theological word the MTA often uses for "transformation."

The MTA is a robust organization with professional leadership and an active schedule of conferences and other programs. Mormon theology calls for a glorified and immortal body achieved through *theosis*, a notion we discussed earlier in this chapter in connection with the Eastern Orthodox branch of Christianity. The MTA contends that Mormon scriptures teach that God commands we use science and technology to help achieve this theological goal. Here is the MTA list of affirmations:

[36] Tracy J. Trothen, "Technology, Medicine, Ethics and Religion: Body Matters," in *Bloomsbury Religion in North America* ("Religion, Science and Technology in North America" section, section eds. Whitney Bauman and Lisa Stenmark) (NY: Bloomsbury, in press).

[37] p. 51.

[38] "Extreme Longevity Research: A Progressive Protestant Principle," in *Religion and the Implications of Radical Life Extension,* eds. Derek Maher and Calvin Mercer (New York: Palgrave Macmillan, 2009; republished in paperback, 2014), 58.

[39] https://transfigurism.org/.

- We are disciples of the Gospel of Jesus Christ, which is to trust in, change toward, and fully immerse our bodies and minds in the role of Christ, to become compassionate creators as exemplified and invited by Jesus.
- We believe that scientific knowledge and technological power are among the means ordained of God to enable such exaltation, including realization of diverse prophetic visions of transfiguration, immortality, resurrection, renewal of this world, and the discovery and creation of worlds without end.
- We understand the Gospel to be compatible with and complementary to many religions and philosophies, particularly those that provoke strenuous pursuit of compassionate and creative exaltation.
- We feel a duty to use science and technology according to wisdom and inspiration, to identify and prepare for risks and responsibilities associated with future advances, and to persuade others to do likewise.
- We seek the spiritual and physical exaltation of individuals and their anatomies, as well as communities and their environments, according to their wills, desires, and laws, to the extent they are not oppressive.
- We practice our discipleship when we offer friendship, that all may be many in one; when we receive truth, let it come from whence it may; and when we send relief, consolation, and healing, that raises each other together.

Notice that the affirmations do not require members to be Mormon, and, indeed, a recent quick survey shows that about 40 percent of members do not self-identify as Mormon.

To further develop the view of this organization, from the MTA website we include a statement that echoes themes from process theology, considered earlier in this chapter.

> Mormonism maintains that God wasn't always God, but became God, and that humans should try to become gods too. Over time, we can become like God by taking steps to improve ourselves and our world … *Mormon Transhumanism* takes the Mormon idea that humans should become gods, and the Transhumanist idea that we should use science and technology in ethical ways to improve our condition until we attain posthumanity, and suggests that these are related, if not identical tasks. That is, we should ethically use our resources including religion, science, and technology to improve ourselves and our world until we become Gods ourselves.

Lincoln Cannon is a founder, board member, and former president of the MTA. In "Mormonism Mandates Transhumanism,"[40] Cannon argues just that. Cannon proposes and supports four points: (1) God commands us to use pre-

[40] In *Religion and Human Enhancement: Death, Values, and Morality*, eds. Tracy J. Trothen and Calvin Mercer, in *Palgrave Studies in the Future of Humanity and Its Successors*, series eds. Calvin Mercer and Steve Fuller (New York: Palgrave Macmillan, 2017; reissued in paperback, 2019).

scribed means to participate in God's work; (2) Science and technology are among the means prescribed by God; (3) God's work is to help each other attain Godhood; and (4) An essential attribute of Godhood is a glorified immortal body. His introduction to Mormonism, transhumanism, and the connection between the two is also found in a video entitled "Mormon Transhumanism."[41]

Christian Transhumanist Association

The other group that has emerged as playing a very important role in educating about and advocating for Christian transhumanism is, aptly named, the Christian Transhumanist Association (CTA).[42] The CTA affirmation, while pro-enhancement, is modest enough to be inclusive of Christians ranging from ardent advocates to more cautious proponents. It reads[43]:

- **We believe that God's mission involves the transformation and renewal of creation** including humanity, and that we are called by Christ to participate in that mission: working against illness, hunger, oppression, injustice, and death.
- **We seek growth and progress along every dimension of our humanity:** spiritual, physical, emotional, mental—and at all levels: individual, community, society, world.
- **We recognize science and technology as tangible expressions of our God-given impulse to explore and discover** and as a natural outgrowth of being created in the image of God.
- **We are guided by Jesus' greatest commands** to "Love the Lord your God with all your heart, soul, mind, and strength ... and love your neighbor as yourself."
- **We believe that the intentional use of technology, coupled with following Christ, will empower us to become more human** across the scope of what it means to be creatures in the image of God.

Currently, the CTA is led by founder Micah Redding, who is software developer, writer on the subject of human values and technology, and host of the "Christian Transhumanist Podcast."[44] The CTA held its first national conference in 2018, featuring a presentation by scientist Aubrey de Grey, a leading anti-aging proponent.

We are not aware of any other transhumanist organization associated with a major religion. Roy Jackson, the author of a book on Islam and

[41] https://www.youtube.com/watch?v=AeyJbROo-Pw.

[42] https://www.christiantranshumanism.org/. Of your two authors, Mercer is a member of the Academic Advisory Council of the CTA, and Trothen is a member of the CTA.

[43] The bolded sections are as per the website.

[44] https://www.christiantranshumanism.org/podcast.

transhumanism, ends with a call for a "Muslim Transhumanist Association," for which one suggested affirmation is:

> We believe that the intentional use of technology will empower us to transcend our current state and move towards perfectibility, guided by the example of the Prophet Muhammad as the Perfect Human (*insan al-Kamil*).[45]

The Arabic *insan al-Kamil* is an honorific title for the prophet that means "the person who has reached perfection" or "the complete person." We anticipate that in the next few years transhumanist organizations will emerge in various religions.

Religion and Science

The relationship between religion and science is of increasing interest to scholars of religion and theologians of the various traditions. Many agree that technology and medicine do not give all the answers to every question and that other sources such as religion help us understand what it means to live better and be better. Examine the video, "Can Science Understand Everything?" in which a number of the United States' National Aeronautics and Space Administration (NASA) scientists respond to the question of whether science can provide answers to everything.[46] The level of interest in the relationship between religion and science is such that, in the past few decades, a robust subfield of "religion and science" has emerged with devoted journals, conferences, and professional societies. Simplifying, but hopefully not overly so, three basic models of the relationship between science and religion have been identified: conflict, independence, and dialogue/integration.

That religion and science are contradictory and in conflict through history, at least in the West, was the prevailing view of the history of the relationship between religion and science until recent decades. The seventeenth century Galileo affair was held up as a prime example of the clash between and incompatibility of religion and science. Galileo Galilei (1564–1642) was supposedly punished for his support of heliocentrism, i.e., the belief that the Earth and planets revolve around the sun. A more nuanced interpretation, held by most historians of science and religion today, is that the conflict was more about politics and a dispute about the Roman Catholic Church's authority, rather than an intrinsic conflict between religion and science. Evolutionary biologist Richard Dawkins is a prominent contemporary advocate of the conflict model.[47]

Advocates of the independence model argue that religion and science have different missions and methods and have nothing to do with each other. A

[45] Jackson, *Muslim and Supermuslim*, 175–76.

[46] "Can Science Understand Everything? NASA Scientists Attempt to Answer the Question." https://aeon.co/videos/can-science-understand-everything-nasa-scientists-attempt-to-answer-the-question.

[47] E.g., *The God Delusion* (Boston: Houghton Mifflin, 2006).

common assertion from this camp is that religion has to do with values and science with facts. An example of a proponent of the independence model is evolutionary biologist and historian of science Stephen Jay Gould, who was a respected advocate of what he called the "non-overlapping magisterials" of religion and science.[48] Proponents of the independence model have been criticized for being vulnerable to scientism, which conceals arguments as unchanging "objective facts." As the history of science demonstrates, no definition or argument is infallible. Science, by definition, requires openness to the possibility of being proven wrong.[49]

A dialogue/integration model can have many expressions, depending on the religion, theology, and interests of the advocate of the model. An interesting proponent of this model from an evangelical (i.e., conservative) Christian perspective is Francis Collins, a widely respected scientist who headed up the important Human Genome Project, completed in 2003. Collins founded the organization BioLogos, which illustrates the dialogue/integration model in a thoroughgoing way.[50] Theological ethicist Ron Cole-Turner also promotes the dialogue/integration model, which he articulates from a progressive Christian perspective. Cole-Turner explains how understanding and meaningfully exploring what it means to be human, why we exist, and what is our purpose and destiny require science and religion working together.[51]

The Theological Continuum

A helpful way to think about the three religion and science models (conflict, independence, and dialogue/integration) is to filter them through the liberal-conservative theological continuum presented earlier in this chapter. In general, liberals—human centered, this-worldly, revisionist, pragmatic—are more likely to embrace a non-conflict model of the relationship between religion and science. Usually, conservatives—God-centered, otherworldly, traditionalist, dogmatic—are more prone than liberals to adopt some version of a conflict model.[52] This statement is a generalization and, of course, there are exceptions. We have already noted how evangelical Christian Francis Collins does not advocate a conflict model.

[48] *Rocks of Ages: Science and Religion in the Fullness of Life* (New York: Ballantine Books, 1999).

[49] Jim Parry, "Must Scientists think Philosophically about Science?" in *Philosophy and the Sciences of Exercise, Health and Sport: Critical Perspectives on Research Methods*, ed. Mike McNamee (New York: Routledge, 2005), 22.

[50] www.BioLogos.org.

[51] Ronald Cole-Turner, *The End of Adam and Eve: Theology and the Science of Human Origins* (TheologyPlus Publishing, 2016). Brent Waters provides a cautious approach to theological embrace of technology. See *From Human to Posthuman: Christian Theology and Technology in a Postmodern World*, in *Ashgate Science and Religion Series,* series eds. Roger Trigg and J. Wentzel van Huyssteen (Burlington, VT: Ashgate Publishing, 2006).

[52] An excellent social science presentation of what people of faith think about science is Elaine Howard Ecklund and Christopher P. Scheitle, *Religion* vs. *Science: What Religious People Really Think* (New York: Oxford University, 2018).

Liberals and conservatives often congregate in their respective positions on wide-ranging issues, such as politics, judicial philosophy, economics, and religion. In the coming chapters, however, we will predict that the new world of radical human enhancement may produce some strange bedfellows, with traditional liberal and conservative groupings finding new expressions.

Our reflections up to this point on the relationship between religion and science have been drawn from Christianity, because it is that tradition in which the most vigorous disagreements are taking place. As we have seen, the model of the relationship between religion and science adopted by Christians and Christian organizations is influenced, in large part, by where they lay on the liberal-conservative continuum. In comparison, religions originating in Asia, such as Hinduism and Buddhism, are generally accepting of science, seeing no conflict between religion and science.

In Hinduism, the dividing line between religion and science is blurred in the sense that all of reality is, in the deepest sense, divine, and so an embrace of any way of more fully understanding reality is a spiritual enterprise. Based on teachings stemming from its founder, Buddhism is an empirical religion, encouraging direct, personal experience as the final test for truth. One could even call Buddhism scientific in that it aims at uncovering the causes and effects that order existence, especially the cause and effect of suffering. While Confucianism has a more checkered history with science, Jainism, Sikhism, and Daoism in the main are quite compatible with scientific findings.

Science and Technology

The relationship between science and technology is complex. One of your authors, Professor Trothen, explains the relationship between science and technology in another book:

> … the goal of science is knowledge and technology is the practical application of science. There has been much debate concerning the conflation of science and technology; proponents of one side do not wish to be subsumed by the other. Science and technology are connected, but they are not the same.
>
> One distinction that is often—but not always—made between science and technology concerns values: science is purported by some to be objective or value-free, while technology is generally accepted as value-laden (Frey 2011). Several European philosophers, including Herbert Marcuse (1964), Jürgen Habermas (1971), and Michel Foucault (1988), have shown that technology promotes values of efficiency and utility …
>
> While it is clear that technology is informed by particular values, I am in agreement with American philosopher, historian, and physicist Thomas Kuhn and do not find science's purported objectivity convincing … Scientific knowledge, it has been said, is simply knowledge; the choices around how to apply it have not been made in the science itself. However, factors influencing the framing of science are value-laden and so have an effect on the science: How we choose which scientific

inquiries to investigate and who should be involved in these pursuits are value-laden decisions.[53]

Science and technology are human pursuits and creations and, as such, have a reciprocal relationship with context, including the people who interact with science and technology. For example, the COVID-19 pandemic, some governments placed a strong emphasis on protection for healthcare professionals and consequently invested huge resources in the creation and manufacturing of personal protective equipment (PPE). Other governments emphasized personal freedom or the economy, rather than preservation of life, which led to proportionately fewer resources funneled into healthcare. Our values go far in shaping how political powers direct science and technology.

Concluding Reflection

Our goal in this chapter has been to set the context for a religious consideration of various radical human enhancements and other technological programs in the coming chapters. Radical enhancement is in the context of a growing transhumanist movement that almost assuredly will lead to radical human enhancement to one degree or another and maybe even lead to posthuman beings.

The religions will be impacted by transhuman and maybe posthuman beings, and the religions also will have the opportunity to assess and influence the coming developments. Religion helps shape values and moral reasoning for many people in the world, and, as a result, religion is often embedded in responses to radical enhancement. To put it another way, with regard to the topic of this textbook, academic theologians and lay adherents of religion can and should have much to say about human enhancement technology in the public square.[54]

As we have emphasized, this chapter by no means attempts a systematic introduction to the world's religions. Rather, we have identified some themes in those religions that are being and could be employed in the conversation about enhancement. As noted, Judaism and Christianity have been the most active in thinking about radical enhancement. We want to play a part in expanding the conversation, and so have endeavored to raise questions and issues from religions beyond Judaism and Christianity. In the coming years, scholars, theologians, and followers of all the religions will come increasingly into this important conversation.

[53] Tracy J. Trothen, *Winning the Race? Religion, Hope, and Reshaping the Sport Enhancement Debate* (Macon, GA: Mercer University Press, 2015), 25–26.

[54] See, e.g., Peter Kahn, "Bioethics, Religion, and Public Policy: Intersections, Interactions, and Solutions," *Journal of Religion and Health* 55, no. 5 (2016): 1546–1560; and H. Brody and A. Macdonald, "Religion and Bioethics: Toward an Expanded Understanding," *Theoretical Medicine and Bioethics* 34 (2013):133–145.

Questions for Discussion

1. Is transhumanism a religious or secular concept? Why?
2. What is the difference between a transhuman and a posthuman?
3. After reading Max More's "Letter to Mother Nature," how would you describe More's transhumanist attitude toward nature and our current human capabilities?
4. Why do you think the Abrahamic religions of Judaism and Christianity have been most active in discussions about radical human enhancement?
5. If human beings begin to live for thousands of years, how might the doctrines of karma and reincarnation be revised or reinterpreted to be relevant?
6. What are ways the concept of change or process, as taught by process theology, can be employed to embrace transhuman and posthuman beings?
7. Do you think human beings are basically good or evil? Did you arrive at your conclusion though a religious or some other avenue? Explain.
8. What differences and similarities do you see between the MTA and the CTA affirmations?

CHAPTER 4

Radical Human Enhancement and Ethics: Questions We Must Ask

Good Ethics: Seeing the Complications

Throughout the textbook, we will ask ethical questions about human enhancement. It may seem that human enhancement ethics is about one thing only—whether or not to choose particular enhancements. But many more serious questions loom. Should we proceed with radical human enhancement research? Who should design the enhancements? Who profits from these technologies? How are they marketed? Who has access? How might these technologies change our collective way of being? Each enhancement needs to be explored, and ethical responses may vary enhancement to enhancement. One may, for example, support radical cognitive enhancement but not superlongevity.

Ethics is about more than believing something is right or wrong. Ethics also goes beyond feeling strongly about an issue. Passion is important to motivate and engage us, but passion alone is inadequate for sound ethical analyses. We need to step back from our feelings and understand the origin of our opinions. We need to obtain good information, engage diverse perspectives, and analyze an issue in light of ethical theory and stated values. Good ethics requires passion and reason. Good ethics is complicated. In this chapter, we give you some of the tools to do good ethics.

Perspective and Community

As Professor Trothen tells her classes, by the end of an ethics course students should have more questions and, perhaps, have less easy clarity on issues. Exposure to differing views and experiences of classmates and of various scholarly analyses complicates things, but in a healthy way. By engaging diverse perspectives, questions arise that may not have occurred otherwise.

Perspective is essential to good ethical analysis. Perhaps you have heard of the parable of the blind men and elephant, an ancient and widely circulated

C. Mercer, T. J. Trothen, *Religion and the Technological Future*,
https://doi.org/10.1007/978-3-030-62359-3_4

story from the karmic religions. Each blind man touches the elephant in a different place (e.g., tusk, leg, trunk, stomach), and then each man describes the elephant—inaccurately—based on their limited, subjective experience. You can effectively illustrate this in your own classroom or home. Cover all parts of a chair with sticky notes, and then count the notes, while standing at different angles to, and distances from, the chair. Good ethics depends on understanding we are perspectival and addressing how we know what we know, a philosophy subfield called epistemology.

We must do ethics in community if we are to overcome a limited view of an issue. Good ethics is done consultatively, constantly asking what voices are missing and why. We all are contextualized, and understanding various contexts helps broaden perspective on an issue. Contexts, as we will see, include religious and spiritual beliefs.

Self-awareness and Values

Good ethics requires self-awareness. Consider the influences on your values. Values are those things that are most important to us, such as achievement, success, family, friends, relationships, health, caring, social justice, pleasure, and adventure. Values, beliefs, and opinions are influenced by the "sources of authority" in our lives, such as family of origin, media (including social media), research, education, law, political leaders, faith communities, history, personal experiences, and what some call intuition, inner spirit, or gut sense. Our values can change over time and through different contexts as we meet people, gain insights, and accumulate varied experiences. Sources of authority shape and reshape our values throughout our lives.

Here is a short exercise that helps uncover the role of values in our lives. What are your top five values? Do not overthink the question. As a prompt, consider what would you like written in your obituary. That might sound a bit macabre, but the question can help distill what is most important to you. We are told what we should desire by the news media, social media, livestreaming, movies, political leaders, and even families and friends. Sometimes it can be difficult to get to the bottom of what we really desire and value.

Perhaps modern conveniences and technologies are high on your top five list of values. Technology itself is value laden. Well respected European philosophers have shown that technology promotes values of efficiency and utility.[1] A dishwasher cleaning our pots and pans, quietly and quickly, adds to our desire to acquire more technologies that get things done easily and efficiently for us. Much of what we are "supposed" to desire is shaped by consumerism in a capitalist context. People line up overnight for the newest

[1] Michel Foucault, *Technologies of the Self* (Boston: University of Massachusetts Press, 1988); Jürgen Habermas, *Knowledge and Human Interest* (Boston: Beacon Press, 1971); and Herbert Marcuse, *One Dimensional Man: Studies in the Ideology of Advanced Industrial Society* (Boston: Beacon Press, 1964).

iPhone, even if they do not really need it and even if it will not make their lives better. Getting to what it is that we truly want is an important part of doing ethics well, but it is not an easy matter.

Ask someone else to do the same exercise, and compare your values. Consider why you might share some values and why other values may differ. Do you always choose and act in ways aligned with your top five values? Make a record of your answers, and later in the textbook we will ask you to think about your values and how they might relate to enhancement options and possibilities.

Self-reflexivity

In the service of good ethical analysis, self-awareness leads to self-reflexivity, described well in the following quote. Self-reflexivity is

> the process of reflecting on one's own story from multiple diverging standpoints in ways that try to take into account one's own experience of privilege and disadvantage within intersecting social systems like sexism, racism, heterosexism, and religious forms of oppression."[2]

These systemic power patterns affect what we experience in the world and teach us what we can expect. Sadly, not everyone is treated equally, and we are not all valued as worthy. An intersectional ethic prioritizes and brings together justice concerns, emphasizing the moral relevance of systemic privileges and barriers, as expressed in this description:

> The further [one is] from the norm, the greater the marginalization. This marginalization, however, is not simply additive, but rather social categories of gender, race, class, and other forms of difference interact with and shape one another within interconnected systems of oppression. These systems of oppression—sexism, racism, colonialism, classism, ableism, nativism, and ageism—work within social institutions such as education, work, religion, and the family ... to structure our experiences and relationships in such a way that we participate in reproducing dominance and subordination without even realizing it."[3]

In this textbook we aim to stimulate thinking about how these systemic power imbalances may affect the evaluation of radical human enhancements and biohacking processes. Hopefully, many questions will begin to arise. For example, how might one's views about the elderly be impacted by promotion of anti-aging technologies? What difference does it make if AI algorithms are

[2] Carrie Doehring, *The Practice of Pastoral Care – A Postmodern Approach* (Louisville, KY: Westminster/John Knox Press, revised and expanded edition, 2015), 191.

[3] Kim, Grace Ji-Sun; and Susan M. Shaw, "Intersectional Theology: A Prophetic Call for Change," *Huffington Post* (March 31, 2017). http://www.huffingtonpost.com/entry/intersectional-theology-a-prophetic-call-forchange_us_58dd823de4b0fa4c09598794. The term intersectionality was coined in 1989 by Columbia and UCLA professor Kimberlé Crenshaw.

programmed mainly by white appearing[4] men? Our hope is that you will consider how systems of oppression may be morally relevant to the intersection of religion and radical human enhancement.

Theories of Ethics

To recap, good ethics requires passion and reason, community consultation, self-awareness, and self-reflexivity. Now, we place the ethical project in the context of some leading theories of ethics. Drawing on values, principles, potential consequences, and virtue, these theories provide systematic bases from which to better understand moral and ethical issues and to help us with ethical decision-making. While the two words are sometimes used interchangeably, morals are about our personal character and beliefs about right and wrong, and ethics addresses the accepted rules, actions, and behaviors in a community or group.

Of the numerous ethical theories, we consider three commonly agreed upon core theories: deontological, teleological, and virtue ethics. Two of these theories—deontological and teleological—are decisionist, i.e., focused on how we make decisions in response to ethical questions. Decisionist models ask, "What shall I do?" Virtue ethics theories have a different focus in that they emphasize who we are as individuals and what values we hold. Virtue ethics has us ask, "What character should I possess as a person?" Our character, according to virtue ethics, drives conscientious behavior. We now look more closely at the three theories. In this textbook we pay most attention to virtue ethics and often ask if these technologies will make us better people and the world a better place.

Deontological

"Principlist ethics" is a term sometimes used for deontological ethics. This approach is a normative ethical theory that says that moral behavior should be based mainly on principles. Examples of principles are beneficence (duty to do good), non-maleficence (duty to do no harm), respect for autonomy (includes respect for individual choice and dignity), justice, veracity (truth-telling), fidelity, and self-care. Principles are derived from those virtues and values that we see as being most important. Principles are codes of behavior. Prima facie principles are those principles that are understood as most obvious and universally applicable.

Deontological theories hold principles as the most important guides for decision-making. Most ethicists understand principles as binding but not absolute, meaning that the principles must almost always be followed, except when they come into conflict with each other. For example, in cases where we cannot absolutely follow two prima facie principles, we might try to follow

[4] To say "white" carries with it the faulty assumption that all people who appear white are of one normative racial background.

both to a degree. To illustrate, if someone's leg is gangrenous and the only treatment to save their life is the amputation of their leg, then the harm of cutting their leg off would likely be judged less severe than the harm of allowing them to die. Following the principle of totality, the duty to do good by saving a life outweighs the harm done by removing a leg.

The religions uphold principles that provide a moral compass or a way to live and behave, based on values of the religion. A key principle in Christianity, for example, is the instruction to love one's neighbor as oneself. This duty to do unto others as we would have others do unto us is known as the "golden rule." Religions are concerned with principles, such as the golden rule, the potential consequences of our behaviors, and with being a good or virtuous person.

To illustrate how deontological ethics can be applied to religion, one could hold the theological principle that human bodies are God's temple and, as such, the body should not be changed in any way that cannot be clearly understood as protecting and preserving the embodied person. This theological position could also be related to the principle that we should respect autonomy, which includes each person's dignity. So, this theological conviction (i.e., principle) that bodies are God's temple could have implications for a variety of ethical decisions regarding human enhancement technology.

Teleological

Teleological theories emphasize the importance of possible and anticipated consequences in ethical decision-making. "Telos" is a Greek word meaning "end" or "goal." Teleological or consequentialist theories are usually thought to include situationalism and utilitarianism.

Situationalism holds that all moral decisions are particular to the specific situation; there are no overriding norms. Each situation must be understood apart from preconceived conclusions or rules. However, even situationalists usually acknowledge the one overarching rule that love must be maximized.[5]

Utilitarian theories are also teleological. These theories look to maximize the greatest good for the greatest number. Utilitarians see actions as morally right or wrong depending on the effects of those actions. Utilitarians agree that the overall aim in evaluating actions should be to create the best results possible, but they differ about how to do that. Some utilitarians, who are called act utilitarians, focus on the effects of individual actions and take a case-by-case approach that evaluates the effects of specific actions in specific cases. Other utilitarians, called rule utilitarians, focus on general rules as generally causing certain effects. For example, rule utilitarians might justify limited funding of superlongevity medicines on the basis that these medicines would generally minimize age-related suffering and reduce the health care costs incurred by aging, but would likely prioritize access to basic needs sustaining the majority

[5] See Joseph Fletcher, *Situation Ethics: The New Morality* (Louisville, KY: Westminster John Knox Press,1966/1997).

of people if a choice needed to be made between basic needs and superlongevity interventions. An act utilitarian would assess the use of resources to extend life on a case-by-case basis, depending on the likely effects of extending someone's life. For instance, an act utilitarian might judge that a person who is making dramatic contributions to science and politics should receive expensive life-extending technologies, because their ongoing work will benefit a great number of people.[6] As with other consequentialists, both act and rule utilitarians are most concerned with the effects of choices and actions.

Hedonism is an example of another teleological theory. Hedonism is interested in maximizing pleasure. In short, whatever results in the most pleasure for any person or group is warranted, without regard for other potential consequences.

Critics of utilitarianism, such as liberation theologians, argue that good outcomes for the greatest number can neglect people at the margins who are socially less powerful and often invisible. Also, the assumption that the many are more important than the few conflicts with religious principles that uphold the incomparable value of each life.

Although limited, a utilitarian approach could be helpful, for example, in deciding who got a respirator or dedicated nursing care during the 2020–21 worldwide COVID-19 health crisis, in hospitals that had too limited a supply of equipment or healthcare professionals. When there is only one respirator and two lives are at stake, who gets the respirator? If a decision is not made, and there are no alternatives, such as jury rigging a stop-gap machine or borrowing from another hospital, both people may die. A utilitarian ethics approach may have us consider such factors as who would likely live longer if they survive the virus. On the other hand, as a critique, if numerous similar decisions are made on the basis of who might live longer, the world may lose most of its elders and the experience and possible wisdom elders bring. Or, people with disabilities may die disproportionately, reflecting and amplifying a devaluing of people with disabilities. At the same time, a utilitarian approach can help us to identify relevant factors in rationing situations where difficult choices need to be made. For example, a utilitarian approach would consider the relative likelihood of survival of patients who need a scarce respirator, prioritizing those most likely to survive.

Most of the time, ethicists consider both principles and possible consequences, but some ethicists emphasize principles and others emphasize consequences. In other words, a hybrid approach is common. A virtue ethics approach, discussed below, may also influence how ethical issues are evaluated.

Virtue Ethics

Virtue ethics asks not what should I do (based on principles or consequences) but what sort of person should I be. Highlighting character, the key question is, "What would the most virtuous person we can think of do in a similar

[6] James Rachels, *Elements of Moral Philosophy*, 4th ed. (Boston: McGraw-Hill, 2003), 96-121.

situation?" For Christians, this person might be Jesus. For Buddhists, the virtuous person could be the Buddha. Some people think beyond the iconic "founders" of religions to great saints or other figures, such as Mother Theresa, Mahatma Gandhi, Nelson Mandela, or even Oprah Winfrey and Michelle Obama. Others think of someone more personal, though not necessarily famous, who has made a big impact on their life. Virtues are qualities we deem to be morally good or desirable in people and might include prudence, self-control, generosity, and kindness.

Our values inform the virtues we think of as desirable. As discussed earlier, values are about personal and subjective beliefs, attitudes, and ideals that influence our everyday living. Values are internal for each person and are likely to change over time, more than principles, which tend to be constant and universal. Things most important to us often become embedded in our character as virtues and inform how we behave in various situations (at least we hope so!).

An ethic of care is an example of a virtue ethics theory promoted by many feminists. What would be the most caring way to be in a given situation, is the ethic of care question. For example, if someone thinks they would be successful if they could think more quickly, perhaps it would be caring to provide them with cognitive enhancement that speeds up their thought processes. But maybe it would be more caring to assist this person in exploring why they think they would be more successful with cognitive enhancement. Maybe they have not weighed carefully the possible benefits and harms. A more caring response might be to consider possible implications for all affected by this decision. A feminist ethic of care considers these contextual and relational factors before deciding on a course of action.

A significant challenge faced by virtue ethics theorists is the need to recognize that a response one person judges virtuous may not be the best or most virtuous response for others. Community, including faith communities, and accountability are necessary to all ethics theories, perhaps especially to virtue ethics. The question of what makes us better people is complex. As we get more and more technological options for changing ourselves, we need to ask what makes us truly better, very deliberately and in community.

Weaving the Theories Together

Many good resources are available for exploring these and other ethics theories in detail.[7] For our purpose in assessing radical human enhancements, it is important to know that while principles, consequences, and virtue are all important in good ethical reasoning, none alone is usually sufficient. Our study, thus far, hopefully will prompt additional questions and attention to various

[7] For example, J. Parks and V. Wike, "Theories and Values in Bioethics," in *Bioethics in a Changing World*, eds. J. Parks and V. Wike (Upper Saddle River, NJ: Prentice Hall, Pearson Education, 2010).

moral principles as we address various human enhancements. For example, we will advocate for the principle of co-design, that is, engaging diverse input into the development of technology, so that many perspectives and needs are addressed. In this textbook we are interested in how the commitments and beliefs of religions might affect how ethical issues associated with enhancement technology are evaluated.

Importantly, these different approaches to ethical reasoning are not strictly aligned with either theologically liberal or conservative positions. Those who adopt more liberal theologies, giving attention to social justice, could very well be most concerned with the potential consequences of ethical decisions. But, the more liberal theologian might also choose a strong principlist approach, emphasizing the duties of justice, co-design, doing good, avoiding harm, and/or respecting autonomy. Finally, they may take an approach more in keeping with a virtue ethic, emphasizing the importance of good character on a collective level.

Similarly, most people of a conservative theological persuasion would likely emphasize principles over consequences, with the principles understood as rules stemming from guidance in their sacred scriptures or revered traditions. However, some conservatives may be virtue ethicists (although they may not use that term), attempting to emulate the Buddha or Jesus. Extreme conservatives, i.e., fundamentalists, are likely to be strongly informed by a literal interpretation of scripture as a source of authority.

We encourage you to remember these theories as you read about ethical issues in each chapter. Consider how we are paying attention to potential consequences, principles, and questions of character.

Three Ways of Defining the Issue

How an issue is defined shapes the moral discourse that follows. The ethical issues surrounding radical human enhancement that intersect with religion tend to be defined in three main ways, as an issue primarily about (1) what it means to make us better individually, (2) morphological freedom of choice, i.e., freedom to make choices about modifying our own bodies, and (3) justice, including social marginalization, resource allocation, and access to relevant technologies.

The Therapy—Enhancement Continuum: What It Means to Make Us Better

A common approach to the ethics of radical human enhancement is to frame the issue in terms of where the enhancing technology might fall along a continuum with therapy at one end and enhancement at the other end. Therapeutic interventions are those interventions that bring us to or keep us at normal functioning; enhancement takes us beyond whatever we currently think of as normal human functioning. A simplistic view is that therapy is acceptable, and enhancement is unacceptable. But, of course, once we begin asking good questions, things get complicated, because some therapies may be unacceptable for various reasons (e.g., risky side effects or because they are contrary to a religious conviction) and some enhancements may be acceptable (e.g., for our purposes in the textbook, because they are consistent with religious belief). Also, it is not always clear if a particular intervention is a therapy or an enhancement. In fact, it may be both, depending on how one views it.

If an intervention is seen as therapeutic, it is, all other things being equal (an important qualification), ethically permissible, and possibly even ethically mandatory. For example, a pacemaker needed to regulate someone's heart rate is a therapeutic technology in that it helps keep the person alive. However, a pacemaker is arguably also an enhancing medical intervention in that it prolongs life and improves quality of life.

Earlier in this chapter, we discussed the COVID-19 ventilator, which in one sense is a human enhancement technology in that a machine is providing someone with life extension. Both pacemakers and ventilators, however, have come to be seen as normal and acceptable over time. No one usually questions the ethics of using pacemakers and ventilators, but more radical interventions are questioned, as we shall see. These two examples illustrate the therapy end of the continuum and introduce us to the in-between zone of the continuum in which some technologies are both therapeutic and enhancing.

Some technologies that fall in the in-between zone on the therapy—enhancement continuum may be seen as ethically permissible but not ethically mandatory. For example, consider laser therapy that improves vision to slightly better than 20-20. Other enhancement interventions are further toward the enhancement side of the continuum and are seen as more questionable in terms of their value and ability to make people better overall. An example is the use of the pharmaceutical modafinil (e.g., Provigil) to improve cognitive functioning in someone who does not have a medical condition impairing their cognition. Another example of an intervention that is far along the continuum as an enhancement is the covert use of anabolic steroids by competitive athletes. Anabolic steroids pose many health risks, are not medically justified, and are used to make one "better" (enhanced) by increasing muscle strength and bulk.

While this distinction between therapy and enhancement is helpful, we are seeing that it can be very complicated. The definition of normal human functioning is debated and so are questions about what is just below or just above normal. Perhaps eyeglasses are therapy; perhaps they are enhancement. Is 20-20 vision normal and, if so, for whom? Major League Baseball (MLB) players average better than 20-20 vision. Their normal, then, is different from the general population. So, is it therapy or enhancement for MLB players to have laser eye surgery that brings their vision up to the average for MLB players? For the general population, since vision normally begins deteriorating by age 50, are eyeglasses and laser surgery, both of which bring 20/20 vision, enhancement or therapy?

Examples abound. A nursing home resident uses a wheelchair to locomote about the grounds. A wheelchair could be designed to move her twice as fast as she could walk normally. That would be enhancement, as opposed to therapy.

Perhaps it is morally unacceptable to take her beyond what would be normal. Or, maybe increasing her mobility would allow expression of her humanness more fully. We are reminded of an interesting book title, *Rebuilt: How Becoming Part Computer Made Me More Human.*[8]

Most ethicists agree a grey zone exists between interventions that are clearly therapeutic and those that are clearly enhancing. Interventions falling in this in-between space have been variously called restorative,[9] preventative,[10] or non-therapy.[11] In the future, when AI-fueled technologies and other radical therapies become normalized, it will be important to continue asking the question of where on the continuum should we locate a once-radical, now normal intervention.

Most enhancements addressed in this textbook are radical, clearly taking bodies and minds beyond what is generally—and currently—considered normal. Transhumanists, however, usually push back from the long perspective of evolution, arguing it is normal for humans to use technology to extend physical, cognitive, and emotional capacities. Humans have been improving themselves since cavepeople figured how to use a twig to manipulate food into their mouths. Much more recently, our grandparents used adding machines to more quickly and accurately manipulate numbers. Surely, adding machines enhanced cognitive capacity. Today, almost everyone uses computers to extend cognitive capacities in memory, information processing, and communication. These are normal evolutionary advances, says the transhumanist. Perhaps it is actually *abnormal* to halt the process of improving ourselves.

Religion clarifies, and perhaps complicates, the therapy—enhancement continuum. Medical therapies traditionally are applied to bring one's health to a normal range. Religion, generally, may go further in the sense of promoting a more holistic well-being by supporting healing and the lessening of unnecessary suffering. Buddhists believe suffering is reduced by releasing one's attachment to the self and letting go of craving. While suffering, in Hinduism, is understood broadly in terms of the unfolding of the law of karma, the religion also teaches compassion and the value of relieving unnecessary suffering. Daoism advocates practices that nurture health and long life. Christians look to stories of Jesus' healings. In Judaism, the preservation of life (Hebrew, *Pikuach nefesh*) is the highest mitzvah.

[8] Michael Chorost, *Rebuilt: How Becoming Part Computer Made Me More Human* (New York: Houghton Mifflin, 2005).

[9] See, for example, Rob Beamish, *Steroids: A New Look at Performance-Enhancing Drugs* (Santa Barbara, CA: Praeger, 2011), 63.

[10] Preventative interventions are those designed to retain a "normal" state. See Ronald Green, *Babies By Design: The Ethics of Genetic Choice* (New Haven, CT: Yale University Press, 2007), 60.

[11] Andy Miah, "Towards the Transhuman Athlete: Therapy, Non-Therapy and Enhancement," *Sport in Society* 13, no. 2 (2010): 221–33.

The concept of normal is related to the question of what it means to be human, and religion speaks to this question. As the typical dividing point between acceptable therapies and unacceptable enhancements, the concept of normal is seen in particular ways when a religious lens is applied. In an unjust world, the meaning of normal and whose normal is accepted (and promoted) is contentious. Religions can help us to consider ontological questions (e.g., questions about being and existence), including what it is that we value about being human and what it is that we might change or improve. It may be that some radical enhancements that clearly fall in the enhancement end of the continuum are desirable or at least congruent with the values and principles of some religions. A secular understanding of normal is not necessarily the dividing point, from a religious perspective, between interventions that are considered acceptable therapies and interventions that are considered unacceptable enhancements.

One of your authors sees the dividing point between an acceptable and unacceptable enhancement, from a Christian perspective, as being at the point that

> enhancement technologies cause us to cease being divine image-bearers… [since Christian doctrine says that humans are created in the *imago Dei*, i.e., image of God.] An intersectional and relational interpretation of the *imago Dei* suggests that this dividing point comes when we cease to value the interdependence of life, and refuse our creaturely responsibilities to use our creative abilities … to enhance relationships and particularly the well-being of the marginalized.[12]

Beyond Christianity, other religions also understand the just care of all life as very important and congruent with their beliefs. The interdependence of life and the safeguarding of life are criteria that contribute to an understanding of the dividing point between morally acceptable and unacceptable technological interventions for the religions.

History shows that new medical interventions, initially seen as shocking or even repugnant, often become accepted, desired, and viewed as normal. Assumptions about normal often say more about the context of one's interpretation than about what is really normal for diverse people. If the acceptable-unacceptable division on the therapy—enhancement continuum is equated with whatever is acceptable and congruent with a religion's values and theology, then a meaningful project is to determine how emerging enhancement technologies fit—or disrupt—religious claims. Disruption is sometimes needed to foster new engagement between theology and technology. Overall, the dividing point between acceptable and unacceptable technology, for the religions, is not the fraught concept of normal. The dividing point between acceptable and unacceptable technology is the point at which the use or design of a technology no longer holds religious integrity; the dividing point is that point at which an intervention violates religious beliefs.

[12] Tracy J. Trothen, "Moral Bioenhancement Through An Intersectional Theo-Ethical Lens: Refocusing on Divine Image-Bearing and Interdependence," *Religions* 8, no. 5 (2017): 7. https://doi.org/10.3390/rel8050084.

Choice

A second way of defining the issue is to see radical human enhancement as primarily about freedom of individual choice and, more generally, human agency. Agency is the capacity to make choices and to act on them. Most transhumanists believe that enhancements should be accessible to all, but not forced on anyone. In our own words, here is what we hear ardent transhumanists saying: "If you don't want to use these technologies, that is certainly your choice, but do not prohibit me from freely taking advantage of means that might give me health and happiness for 500 years."

Choice is not that simple, however. The extent of one's agency depends on social attitudes and structures, such as race, gender, religion, sexuality, age, and socio-economic class. Agency must be bound together with justice. For example, a person who cannot financially afford genetic modification technologies has a greatly reduced capacity to choose one of these enhancements.

Extreme Individualism

An extreme individualistic ideology insists we should get to make our own choices concerning our bodies regardless of any other factors. People have the right to choose an enhancement, no matter how radical. Medically assisted death is a choice, under certain conditions, in 28 countries.[13] The legal freedom to make so many big choices, as individuals, reflects context. In North America, choice and extreme individualism are highly valued. However, making these choices is not as simple as it may seem. Asian culture, which produced the karmic religions, is more balanced on this point, giving more emphasis to community.

Individualism is deeply ingrained in normative Western culture. We do not like feeling dependent on others. We want to believe we can rely on ourselves, alone. Yet, just about everything we do and attain, engages others in some way. Consider one's daily meals and the many people having a role in supplying that food. A simple loaf of bread on the table depends upon farmers, truck drivers, store owners, store employees, and many others. So, seemingly simple choices involve others, and good ethics asks how those choices impact others.

Relational Autonomy: The Interconnection of Life

Religions have much to say about the interconnection of all life. What we choose and what we do affects many other people and the environment. Few things underscored the reality of interdependence more than the 2020–21 COVID-19 pandemic. Partly because of our global interconnectedness, the

[13] E.g., Belgium, Canada, Colombia, India, Israel, Japan, Mexico, Luxembourg, the Netherlands, Switzerland, Turkey, and parts of the United States. See "Euthanasia & Physician-Assisted Suicide (PAS) around the World," *Britannica ProCon* (2/26/20). https://euthanasia.procon.org/euthanasia-physician-assisted-suicide-pas-around-the-world/

virus spread relentlessly. Also, because we are so interconnected, most people recognized the necessity of physical distancing. The world had to grapple with hard questions about the allocation and sometimes rationing of personal protective equipment (PPE). Should one country limit access of other countries to PPE, and what if one country had a greater need for PPE based on the number of COVID-19 cases and their stockpile of PPE? Countries made decisions about re-opening economies while weighing many issues, such as economic hardship, the mental health impact of physical distancing, and the possible subsequent waves of COVID-19, on the world. Similarly, climate change is a pronounced expression of the interconnection of life. What we do to the environment affects everyone globally and could have disastrous ramifications. On the other hand, as the pandemic also showed us, we can make a collective positive impact on water-life and air quality quite quickly by strongly limiting emissions.

The impact of radical enhancements discussed in later chapters go far beyond the individual being enhanced. For example, a moral bioenhancement that makes one more altruistic leads to being kinder and more self-giving to others. And, not taking the moral bioenhancement will also have implications for others. A runner chooses a physical enhancement, increasing endurance, and that runner becomes more competitive in distance running. Competitors may want to improve their odds against the enhanced runner, so they take the enhancement. And, because the enhanced runner has more endurance, they may spend less time with family.

Respect for autonomy is widely recognized as a key bioethical principle. How autonomy is defined, though, has become increasingly debated. Respect for autonomy includes respect for a person's dignity, which is usually understood to include the ability to make choices about oneself. In normative North American culture, an extreme individualism has reduced a popular understanding of autonomy to individual rights and choice. But a relational autonomy understands rights and choice as relational concepts that only have meaning in the context of each individual's life narrative. This life narrative necessarily includes all the effects our choices may have on other people and other life. Asian cultures, to pick one type, are in some ways better positioned than Western ones to make choices and develop behaviors that respect relational autonomy. We will explore the implications of a relational autonomy throughout the chapters of this textbook.

What Do We Really Want?

It may seem evident that we make choices, and we do. However, as discussed earlier, values and desires are influenced by complex social processes. We get input from so many sources about what we should want and how to go about getting what we want. It can be difficult to uncover what it is that we truly desire and to resist seeking what we are told to want. Consider the exercise, earlier in the chapter, where you identified your top five values. Does the latest and greatest iPhone help you live into your top values, giving you deep satisfaction?

We know that much of our daily life is lived with our brain on "autopilot," out of what researchers call the "default mode network." With this "fast brain" we make decisions quickly, propelled by any number of influences, such as social media and advertising. Making choices consistent with our values can require effort and time for critical reflection.[14]

As we introduce radical enhancements in later chapters, resist an unconscious, quick, "autopilot" response and consider how thoughtfully identified values help you assess the true value of each enhancement. Religion, of course, informs values, explicitly for faith adherents and perhaps unconsciously for others.

Justice

A third way of defining the issue is to see radical human enhancement as primarily about justice. The moral relevance of social systemic advantages and disadvantages, resource access, and resource allocation are all issues of justice, a key principle in religion.

The Religions

Karmic religions emphasize the importance of good deeds and compassion; good karma comes back to benefit those who practice and live with kindness and compassion. The Abrahamic religions also understand the work of justice as important. In Judaism, the commandment to love one's neighbor as oneself[15] and the aspirational concept of *Tikkun olam* (to repair the world) are guiding principles. In Christianity, the commandments, "You shall love the Lord your God with all your heart, and with all your soul, and with all your mind,"[16] and "You shall love your neighbor as yourself,"[17] are key passages and are often associated with a preferential option for the poor.

This preferential option means that those on the social margins, including those marginalized on the basis of their religion, are seen as most important to the world's collective work of understanding injustice and working towards alleviating oppression. The preferential option for the poor is a principle embraced in Christian liberation theologies. A concern is that radical

[14] See e.g., Steve Ayan, "The Brain's Autopilot Mechanism Steers Consciousness," *Scientific American* (12/19/18). https://www.scientificamerican.com/article/the-brains-autopilot-mechanism-steers-consciousness/; and Jessica Hamzelou, "Your Autopilot Mode is Real--Now We Know How the Brain Does It," *New Scientist* (10/23/17). https://www.newscientist.com/article/2151137-your-autopilot-mode-is-real-now-we-know-how-the-brain-does-it/#ixzz6YPqnV6DL.

[15] Leviticus 19:8.

[16] Matthew 22:37.

[17] Matthew 22:39.

enhancement technologies will amplify existing injustices unless we first, or at least simultaneously, address the unhealthy values promoting systemic power imbalances. To work towards the safeguarding of marginalized people as we create more enhancing technologies, we need increased engagement of religion in the public square.[18]

Distributive and Procedural Justice

Distributive justice is about the socially just distribution of resources. The resources distributed may be tangible (e.g., food, pay, or technology) or intangible (e.g., encouragement, valuing people by engaging with them in conversation).

Procedural justice is defined as the fairness of the processes that lead to outcomes. Co-design is a procedural justice principle. Questions such as how much time is allocated to whom and who makes decisions are important to procedural justice. When individuals have a voice in the process or if the process involves characteristics such as consistency and fairness, then procedural justice is enhanced. In the healthcare context, procedural justice questions include who receives care, how long must they wait for care, what quality of care do they receive, and what are the roles of the patient, the patient's family, or multidisciplinary healthcare professionals.

Here is an example of the distributive and procedural justice concerns with regard to access to radical enhancement technologies. Polyurethane super swimsuits changed elite swimming competitions by making swimmers more buoyant and more easily propelled through water. Athletes wearing body-length super swimsuits broke 43 records at the 2009 Fédération internationale de natation (FINA) international swimming competition, and then the suits were banned due to their pronounced effect on swimming performance. Since then, other high-technology drag-reducing swimwear has been developed that is permitted in competitions. But these swimsuits cost a lot of money (200 to 600 US dollars) for people with limited resources, and the suits wear out quickly. Most suits last about two swim meets or 40 swims before the material degrades. This is a clear case of financial means impacting access to enhancement technology. On top of this are procedural limitations in that not every athlete had a voice in deciding whether these swimsuits should have been made an option for competitors in the first place. Should the swimsuits have even been developed?

Turning to the medical field for an example, the gene therapy Luxturna treats a certain type of congenital blindness. Obtaining this treatment requires about 850,000 US dollars. Even though curing blindness is considered therapeutic,

[18] H. Brody and A. Macdonald, "Religion and Bioethics: Toward an Expanded Understanding," *Theoretical Medicine and Bioethics* 34 (2013): 133-145; and Peter Kahn, "Bioethics, Religion, and Public Policy: Intersections, interactions, and Solutions," *Journal of Religion and Health* 55, no. 5 (2016): 1546–1560.

rather than enhancing, therapeutic status does not automatically insure access, and not all therapies have therapeutic healthcare status in all countries.

From the liberal side of the theological continuum, an argument is made that scarce resources should not go to enhancement research when so many are starving, homeless, and without education. Resources devoted to one research area are not available for other areas. As we have noted, who is privileged to make such decisions is critical. Currently, most leaders and proponents of transhumanism are white, socioeconomically privileged, Euro-American men. Marginalized groups are not well-represented among transhumanist proponents. And only recently have any ethics guidelines regarding research, development, and application of human enhancement technologies been drawn up. The European SIENNA project (www.sienna-project.eu), "Stakeholder-Informed Ethics for New technologies with high socio-ecoNomic and human rights impAct," "which seeks to develop ethical protocols and codes for human genomics, human enhancement, and artificial intelligence & robotics," have ethics guidelines in progress.

We have made the point that technology, in the hands of human beings with worldviews, is not value neutral. This is where the procedural justice notion of co-design can be valuable. Marginalized groups must a have a voice at the beginning of the design of AI and other technology, or we risk catering to the preferences of those with power. Without diverse input into the funding, direction, and distribution of enhancement technologies, we risk perpetuating and amplifying unjust power structures.

Social Justice

Social justice is a concern for equity, particularly for the socially vulnerable and the socially marginalized. Social justice is about the protection and empowering of those with less power due to racialization, socio-economic disadvantage, ageism, disability, sexism, gender, sexual identity discrimination, and/or other injustices.

Taking the analysis of power to another level, some ethicists and religionists are concerned about ending up with two classes of people, the advantaged enhanced class and the disadvantaged unenhanced class. People with more power and money, at least initially, will almost assuredly have better access to a range of enhancement options. The concern about classes of people can be extended to countries. Globally, countries with big pharma and tech companies producing enhancement therapies and technologies will see their Gross National Product increase and acquire more political power.

Questions related to co-design, also called participatory design, are important procedural justice and social justice issues. Co-design is the move to involve all stake-holders in decision-making processes, including design, regarding the creation of technologies. Technologies that best meet the needs of people and are as usable as possible would be designed and created with the participation of representatives from socially marginalized communities. Without input, for instance, from those with different abilities and experiences we run the risk of ableism, ageism, racialization, and many other forms of discriminatory tunnel-vision.[19]

Without good faith intentions and clear-eyed moral vision, widening gaps between the haves and have-not individuals and countries will persist and escalate. Your top five values, along with plenty of other commendable values, religious and otherwise, will get lost in the shuffle. Justice concerns can and should complicate our thinking.

Precautionary and Proactionary

"Precautionary" and "proactionary" are technical ethical terms that can be helpful as we move through the following chapters. A proactionary approach to enhancements advocates for the development of life-improving and even life-prolonging technologies, in spite of some risks. The term "proactionary" was apparently coined by Max More, an early transhumanist, as a way of countering the prevailing precautionary principle.[20] Transhumanists generally favor a proactionary stance.

A precautionary position involves moving new therapies and technologies along slowly, paying very careful attention to possible unknown and unintended harmful side effects and, above all, doing as little harm as possible. So, applied to radical human enhancement, a precautionary approach errs on the side of safety by not moving forward quickly on an enhancement if the dangers are unknown or suspected. A proactionary stance errs on the side of speedy production of likely beneficial interventions, in spite of some possible harms. The interpretation and weighing of potential harms and benefits is at the crux of choices involving precaution or proaction.[21]

Bioethicist John Harris is a vigorous advocate of the proactionary approach to some human enhancement therapies, as seen in this statement:

> Therapy delayed is therapy denied, and those who oppose the introduction of new therapies that promise to reduce suffering and extend life face a responsibility as grave as do those who would recklessly introduce technologies that might cause more harm than the good expected or hoped of them. The tension between caution and recklessness walks both sides of this street."[22]

[19] Brashear, R. P. (director), "Fixed: The Science/Fiction of Human Enhancement," *New Day Films* (2013).

[20] For a history of the term, see Steve Fuller and Veronika Lipińska, *The Proactionary Imperative: A Foundation for Transhumanism* (New York: Palgrave Macmillan, 2014), 12-43. See 29-30 for More's role. For Fuller's proactionary view, see, e.g., Steve Fuller, *Nietzschean Meditations: Untimely Thoughts at the Dawn of the Transhuman Era,* Posthuman Studies 1, ed. Stefan Lorenz Sorgner (Basel: Schwabe Verlag, 2019), 76-80.

[21] An example of a blended approach, attempting to achieve a delicate balance, is Daniel McFee, "The Risks of Transhumanism: Religious Engagements with the Precautionary and Proactionary Principles," in *Religion and Transhumanism: The Unknown Future of Human Enhancement,* eds. Calvin Mercer and Tracy J. Trothen (Westport, CT: Praeger, 2015), 217-28.

[22] John Harris, "How To Welcome New Technologies: Some Comments on the Article by Inmaculada de Melo-Martin," *Cambridge Quarterly of Healthcare Ethics* 26 (2017): 166–172. https://doi.org/10.1017/S0963180116000736.

The duty of beneficence (to do good) can be tougher to follow than the duty of nonmaleficence (to do no harm). If regulatory bodies too quickly approve a therapy, the "therapy" may do more harm than good. But, as Harris points out, if a promising therapy is held back, some people may die unnecessarily if the therapy would have worked. The effort to produce vaccines for the 2020 COVID-19 pandemic is an excellent example of the clash between precautionary and proactionary approaches. Some people are prepared to receive a vaccine that has not gone through all phases that are customarily part of a trial, because they would rather risk an uncertain vaccine than getting the coronavirus. Others take a more precautionary approach and want to wait for a potential vaccine to go through the usual testing rigor before they consent to receive the vaccine.

Religion can weigh in on the side of proaction or precaution. An example of a well-developed proactionary approach as an intellectual foundation for transhumanism, and one that considers the religious dimension of proaction, is titled *The Proactionary Imperative: A Foundation for Transhumanism*.[23] Ted Peters is an important and widely respected voice in the Christian theological and ethical assessment of radical enhancement. Peters is not opposed to radical human enhancement, but he insists on moving carefully and cautiously because of significant capacity to sin and to use enhancement procedures for self-serving purposes.

The Power of Words

Throughout this chapter on ethics and, indeed, throughout the remaining chapters, it is important to use language carefully and with a critical eye. Words are potent in shaping moral discourse. For example, consider the word "progress." The word itself implies positive change.[24] But new technology is not necessarily *de facto* good and does not necessarily result in progress. In a similar vein, when we say "enhancement technologies" or "anti-aging technologies/ medicines," we cast an assessment on the topic just by those word choices.

Another good example is the term chosen to refer to those technologies that extend human life indefinitely. "Prolongevity" and "extreme longevity" are both used in the conversation. The first term connotes a more positive assessment of the technologies than the second. The term "radical life extension" could carry a positive or negative connotation, depending on one's thinking about whether things "radical" are desirable or not. Enhancements make us "better." What is meant by "better," and who gets to decide? Self-awareness

[23] Steve Fuller and Veronika Lipińska, *The Proactionary Imperative: A Foundation for Transhumanism* (New York: Palgrave Macmillan, 2014).

[24] Michael Burdett presents an excellent analysis of the meaning of progress and how this value laden term is used. See "The Religion of Technology: Transhumanism and the Myth of Progress," in *Religion and Transhumanism: The Unknown Future of Human Enhancement*, eds. Calvin Mercer and Tracy J. Trothen (Westport, CT: Praeger, 2015), 131–148.

and self-reflexivity are important if we are to uncover power dynamics behind words and be intentional regarding the values that inform ethical reasoning.

How you think about and define the ethical principles described in this chapter help determine whether or not you are generally supportive of radical human enhancement and what particular enhancements you might encourage. "Supportive," of course, can mean different things. When particular enhancements become available, perhaps you will personally choose to avail yourself, or your children, of them. Perhaps you think government funding and legal regulations should be designed to promote enhancement research and development. No major politician on the world stage is yet talking about radical human enhancement, but perhaps you will support politicians partly on the basis of whether they are for or against these enhancements.

Believing the government should support research on any particular enhancement is related to whether or not one thinks that enhancement is inevitable. Over the past decade or two, as relevant technologies have become more powerful, the general consensus among scientific experts has gradually shifted from whether or not new enhancements will be developed to when they will be and by whom. One view is that if the developed countries prohibit some extreme enhancements, the technology will be created anyway, either by rogue countries or terrorist groups. This "off-shore" possibility can be an argument for responsible countries to proceed with research and development, in order to get there first and maintain control over it.

As we move through the radical human enhancement possibilities in the coming chapters, we will see that society and individuals are going to confront hard choices in the coming years. The three frames of therapy—enhancement continuum, choice, and justice, along with the three ethics theories discussed in this chapter, can provide guidance in reflecting on the various radical possibilities.

Questions for Discussion

1. How does your personal story inform your values and how you think about radical human enhancement?
2. Which of the three ethical theories discussed (deontological, teleological, virtue) best describes your ethical decision-making process?
3. Are contact lenses therapy or enhancement? What about a hearing aid? What about your computer? To be an enhancement, does the technology have to be inside us, or attached to us?
4. What do you consider the most important way of framing the radical enhancement issue (i.e., the therapy—enhancement continuum, choice, or justice)? Discuss why.
5. Should you be able to choose whether to get a tattoo? What about expensive eye surgery? Or a germline modification to improve your strength?

6. Regarding possible medical treatments, are you more proactionary or precautionary?
7. Do you think radical human enhancement will ever become the number one social and political issue? Why or why not? If yes, then when?

PART II

Five Categories of Enhancements

CHAPTER 5

Superlongevity and Other Physical Enhancements

Technology

The story of Siddhartha Gautama, known as the Buddha ("enlightened one") after his awakening, is a compelling religious biography. The "Legend of the Four Passing Sights" is told in different versions. The essence of the story is as follows.

Tradition insists young Gautama had everything. He was extremely handsome, of noble family descent, lived in luxury, and had a model wife. The father, who wanted his son to become a universal monarch, shielded him from the harshness and sorrows of life, knowing that old age, disease, and death can drive one to religion. The gods, however, intervened, exposing Gautama to the severities of life.

In the "Legend of the Four Passing Sights," Gautama ventures from the palace, and from his chariot he sees the first sight of misery, an old man—feeble, body bent, broken-toothed, gray of hair. Puzzled, Gautama turns to his charioteer, who explains that aging and death are the miserable fates of every person. His second sight is a sick man, body riddled with disease. The charioteer explains that all bodies are subject to pain and suffering. Ever more sorrowful, Gautama next spots a body in a funeral procession and learns that the stark reality of death follows old age and disease.

At this point in the story, the troubled young Gautama and modern transhumanists see the same unhappy paths of human aging and mortality, with the modern version focusing on nursing homes, cancer wards, and funeral parlors. For Gautama and transhumanists, the sufferings of aging are repulsive, disgusting, and sap the joy out of living. But here the stories diverge.

The fourth sight appearing before the eyes of Gautama was a calm ascetic, meditating at the edge of the forest. This sight gave Gautama hope; he forsook his life of luxury, exchanged his rich garments for the course yellow monk's robe, and plunged into the forest. After years of intense struggle, he achieved

C. Mercer, T. J. Trothen, *Religion and the Technological Future*, https://doi.org/10.1007/978-3-030-62359-3_5

enlightenment, the end of suffering. Transhumanists are just as committed and work just as hard to end suffering, but they look not to religion. They turn to science and technology for salvation from aging, illness, and mortality.

One way to envision our task in this textbook is to probe whether or not the religious path and the transhumanist path might converge. In the particular case of Gautama, the issue is whether technology can serve as a means to alleviate suffering in a way consistent with the teachings of the Buddha.

Would You Take the Superlongevity Pill?

What if you found out you could live a healthy life lasting 500 years—or longer? Making it concrete, what if there was a pill you could take right now that would have you live healthy indefinitely? You could live unless you died from a horrible accident or something like a meteor hitting the earth. Would you take the "live-just-about-forever" pill?

Consider taking an informal poll of your class, family, or circle of friends. It is an interesting question to work into conversations at social gatherings, as the media increasingly picks up on the ever-expanding superlongevity[1] research programs. It is very important, however, to make clear that no superlongevity proponent is talking about being in a nursing home bed hooked up to tubes for hundreds of years. Transhumanists want a healthy life that includes full functioning in the world.

Over the years, your textbook authors have informally polled their classes and audiences at public lectures on the topic. On one end of the response spectrum is something like, "Where's the glass of water, I'll swallow the pill right now." On the other end of the spectrum are those who are adamant they would never, ever choose to live 500 years, even if they were healthy for all 500 years.

Longevity enthusiast Aubrey de Grey quips that if you ask people, who say they do not want immortality, if they are ready to die tomorrow, they never say yes. Most people, even if they vote one way or another, want more information. Can I reverse the decision, and will I have enough money in later years? For sure, the rather singular question—Would you take the longevity pill?—is a bit unfair, because so many qualifications can impact one's decision.

While informal polls are not scientific surveys, our anecdotal evidence suggests that breaking out males and females can yield quite different results. The majority of males we have informally polled would take the pill, while the majority of females would not. In questioning the females about why they would not want to live indefinitely, the most common response is that they would not want to be present in the world without their friends and family.

If the question is qualified by positing that their friends and family members will also take the pill, then the positive female response rate moves roughly to the level of the positive male response rate. These informal results perhaps

[1] We use "superlongevity," "prolongevity," and "anti-aging" interchangeably.

reflect a greater priority given to relationships among females than males. We need ongoing and specific social science research on these kinds of questions. When that comes, as it surely will, it will be interesting to see how people are processing these developments.

In this chapter we consider the category of radical physical enhancement. Since most radical physical enhancements that are being pursued are directed towards radically extending lifespan, we focus on superlongevity physical enhancements as our primary example. You will notice as the chapters unfold that the enhancement categories overlap and interconnect, as do many of the theological and ethical considerations. Humans are not collections of separate parts but are intraconnected wholes. However, we examine these categories of human enhancement—physical, cognitive, moral, affective, and spiritual—separately, because each category is directed primarily at one aspect of human beings, with implications for the other aspects.

Will There Be a Pill?

It is unlikely to be as simple or easy as taking a pill. More likely, there will be several enhancing interventions available to help extend life and put an end to aging. It is hard to find someone who is more laser focused on terminating aging in the human species than biogerontologist de Grey.[2] De Grey has said that extreme longevity, when it comes, might look something like going down to the local health center periodically and sitting for your longevity-producing infusion. De Grey works on a biological path to longevity that may require rejuvenating the body through periodic cleansing of the cell-damaging and life-oppressing waste materials that accumulate in normal living.

By the way, de Grey is not, according to his definition, a transhumanist, although he is certainly one of the heroes of many transhumanists. He is committed to humans living radically long lives, but not in a way that "transcends" humanity. So, he takes issue with the idea that radical life extension is necessarily, or only, a transhumanist idea.

De Grey's biological path to longevity, which he has called the "boring wet approach,"[3] is one of several paths being explored. Currently, there are no medical interventions that stop or reverse human aging. However, possibilities once solely located in the domain of fantasy are being pursued on scientific grounds. Several ongoing research programs—all controversial and debated among ethicists and other scholars—could impact human longevity in a radical fashion. How likely the transhumanist programs are to unfold is the subject of

[2] De Grey's role in the project to end human aging is discussed in James Michael MacFarlane, *Transhumanism as a New Social Movement: The Techno-Centred Imagination*, in *Palgrave Studies in the Future of Humanity and its Successors*, series eds. Calvin Mercer and Steve Fuller (New York: Palgrave Macmillan, 2020), 94. De Grey's lengthy beard is discussed as part of social movement branding.

[3] De Grey, "Dr. Aubrey de Grey—SENS Research Foundation," Life Extension Advocacy Foundation. https://www.lifespan.io/news/dr-aubrey-de-grey-yuri/.

debate. The more radical projects, such as mind uploading (addressed in a later chapter), may be questionable on scientific grounds. But some transformative developments are on the horizon or at least close enough to merit society's attention to their implications.[4]

While these programs involve highly technical, cutting-edge science, the basic thrust of most of them is understandable to the layperson. Body-part replacement, nanotechnology to rejuvenate the body, tissue generation, cyborgization, cloning, digital immortality, mind uploading, telomere extension, and genetic modification technologies are other paths that may individually or in tandem radically extend human life. At risk of greatly oversimplifying the science, telomeres are caps at the end of DNA strands that protect the chromosomes and become damaged and shortened, thereby hastening death. Regarding genetic modification technologies, while aging as a whole cannot be addressed through one specific gene, there are promising indicators that gene therapies could extend more lives.[5] The project to radically prolong life is, indeed, being pursued.[6]

The nature of the superlongevity does matter, in that it has implications for one's experience. De Grey's biological life extension provides the opportunity for continued experience of embodiment in the same basic way humans have always lived. Digital immortality, mind uploading, and cyborgization offer life experience based in an abstract (i.e., non-embodied, at least in a traditional way) notion of the person. These more radical prolongation projects are discussed in more detail in later chapters.

However it happens, advances in medical and other sciences raise the theoretical possibility that biomedical technology could dramatically prolong or even indefinitely extend healthy human life. A key word here is "healthy." As we noted earlier, it is a misconception to think that prolongevity enthusiasts envision us in compromised physical and cognitive states in care homes. The vision, should it unfold, is to live indefinitely in vital, even youthful, bodies and minds or in digital formats where disease and death are non-issues. Optimistic

[4] Physicist Barry G. Ritchie, for example, says extreme longevity achieved through science is unlikely. See "The (Un)Likelihood of a High-Tech Path to Immortality," in *Building Better Humans? Refocusing the Debate on Transhumanism*, eds. Hava Tirosh-Samuelson and Kenneth L. Mossman, vol. 3 in *Beyond Humanism: Trans- and Posthumanism*, ed. Stefan Lorenz Sorgner (Frankfurt: Peter Lang, 2012). The opposite view, that superlongevity is not only attainable, but likely, can be found in some of the chapters in Immortality Institute, *The Scientific Conquest of Death: Essays on Infinite Lifespans* (Buenos Aires: LibrosEnRed, 2004).

[5] For example, in 2019, the United States National Institute of Aging (NIH) reported, in an article by this name, that "gene therapy shows promise repairing brain tissue damaged by stroke" and maybe improving memory and motor skills beyond the pre-stroke level. See Francis Collins, "Gene Therapy Shows Promise Repairing Brain Tissue Damaged by Stroke," *NIH Director's Blog* (September 24, 2019). https://directorsblog.nih.gov/2019/09/24/gene-therapy-shows-promise-repairing-brain-tissue-damaged-by-stroke/. Earlier in 2019, progress was reported to have been made on gene therapy development for treating cardiovascular disease in mice.

[6] See, e.g., Alison Abbott, "First Hint that Body's 'Biological Age' Can Be Reversed," *Nature* (September 5, 2019). https://www.nature.com/articles/d41586-019-02638-w.

predictions for such healthy longevity envision some significant scientific breakthroughs within the next few decades.

Prolongevity: Mainstreaming and Big Money

The possibility of superlongevity may still seem like science fiction to most people. The conversation, however, is slowly moving to the status of a mainstream discussion, at least among experts on aging. Consider the cover title of a recent issue of the *MIT Technology Review*, published by the very reputable Massachusetts Institute of Technology. The cover title is "Old Age is Over! If You Want It."[7] Billed as "The Longevity Issue," this publication contains articles about breakthrough anti-aging drugs in the pipeline and critiques the concept of "old age." This issue of *MIT Technology Review* was not the first issue of this journal to be devoted to longevity research. In February 2005, the respected journal featured de Grey on the cover, with the title "Live Forever? Aubrey de Grey Thinks He Can Defeat Death. Is He Nuts?"[8]

This is a good place to introduce futurist Ray Kurzweil, a living legend and arguably today's leading transhumanist. An accomplished inventor and computer scientist, in 2012 Google appointed Kurzweil as director of engineering and to run their robot natural language project. A year later, Google launched Calico, a start-up company with the mission "to solve death." This appointment of Kurzweil means that a leading transhumanist has been placed in a key management position in a company with enormous resources to put behind the project of "solving death." As a *Time* front page cover line put it, "The search giant is launching a venture to extend the human life span. That would be crazy—if it weren't Google."[9]

Kurzweil uses a computer analogy, asserting that in the past we have depended on hardware (i.e., our biological bodies) as the host for our software (i.e., storage and processing of information in our brain). In the future, he says, we will become just software, no longer dependent on the flesh and blood body. Notice that Kurzweil, as is common in conversations about enhancement, uses analogies that reflect the prevailing technology of the day. In the nineteenth century, minds and brains were described as pipes, cylinders, valves, pistons, and pressure, which are all parts of the then common steam engine. All analogies fall short. The computer analogy is pretty good but is deficient in that a computer does not have a "mind." Just raising this point gets us into interesting and complicated questions, which we will discuss in a later chapter on superintelligence.

[7] "Old Age is Over! If You Want It," *MIT Technology Review* 122, no. 5 (September/October 2019).

[8] "Live Forever? Aubrey de Grey Thinks He Can Defeat Death. Is He Nuts?" *MIT Technology Review* 108, no. 2 (February 2005).

[9] "Can Google Solve Death?" *Time* (September 30, 2013).

Kurzweil was not Google's first reach into radical life extension. As early as 2009, Google appointed another avid transhumanist, Bill Maris, to head the Google Ventures investment fund. Maris said, "If you ask me today is it possible to live to be 500, the answer is yes." Google Ventures has been investing about one-third of its two-billion-dollar portfolio in life sciences start-ups, including ambitious life-extending projects. Maris explained, "We aren't trying to gain a few yards. We are trying to win the game."[10]

One reason we are giving some background regarding Google's involvement in superlongevity projects is to make the point that serious money has been moving into making radical life extension a reality. Several billionaires support life extension research, including Larry Ellison, Paul F. Glenn, Domitry Itskov, and Sergery Brin.[11] As might be expected, Silicon Valley is deep into this quest.[12]

Here is one more example of Silicon Valley interest and money. PayPal co-founder Peter Thiel, whose private fortune has been estimated at 2.2 billion, has stated publicly that he aims to live forever. "I think there are probably three main modes of approach to death," he explained. "You can accept it, you can deny it, or you can fight it."[13]

Funding for technologies that can lead to superlongevity and other radical enhancements comes not just from investors and corporate coffers. Almost everyone who pays taxes, for example, supports, perhaps unknowingly, research that contributes to radical enhancement. In the United States, for example, the Pentagon's Defense Advanced Research Projects Agency (DARPA) and its 3.5-billion-dollar annual budget[14] is heavily involved in building a more effective military through, in part, enhancing soldiers. Enhancing strength, capacity for quick repair of injuries, and other general physical traits can be supportive of superlongevity as well.

The current and anticipated generous funding for superlongevity research has implications. Later in this chapter, we consider justice issues related to this funneling of resources and the power of private technology companies.

[10] Katrina Brooker, "Google Ventures and the Search for Immortality," *Bloomberg* (March 9, 2015). www.bloomberg.com/news/articles/2015-03-09/google-ventures-bill-maris-investing-in-idea-of-living-to-500.

[11] Adam Leith Gollner, "The Immortality Financiers: The Billionaires Who Want to Live Forever," *The Daily Beast* (August 20, 2013). http://www.thedailybeast.com/articles/2013/08/20/the-immortality-financiers-the-billionaires-who-want-to-live-forever.html.

[12] Tad Friend, "The God Pill--Silicon's Valley's Quest to Live Forever: Can Billions of Dollars' Worth of High-Tech Research Succeed in Making Death Optional?" *The New Yorker* (April 3, 2017).

[13] Mick Brown, "The Billionaire Tech Entrepreneur on a Mission to Defeat Death," *The Telegraph* (September 19, 2020).

[14] In fiscal year 2020.

Longevity Escape Velocity

"Longevity escape velocity" (LEV) is a concept coined by David Gobel, co-founder with de Grey of the Methuselah Foundation, a nonprofit organization with the mission to "make 90 the new 50 by 2030" through life extension science. LEV has been championed by de Grey and Kurzweil. Also called "actuarial escape velocity," LEV refers to a coming future time in which we have the know-how to extend our lives long enough for the next scientific breakthrough to come, which will again extend our life expectancy for the next breakthrough. The title to this book by Ray Kurzweil encapsulates the LEV idea, *Fantastic Voyage: Live Long Enough to Live Forever.*[15]

In the modern period, life expectancy has gradually increased each year due to advancements in traditional medicine. However, now it takes more than one year of research to gain a year in life expectancy. When the research catches up to the rate of aging, and this catching up can be sustained, then we have escaped, that is, outrun, death. That is LEV.

Like many radical enhancements, superlongevity is likely not going to be as clear and simple as a pill suddenly appearing on the market that can add 500 healthy years to your life. Rather, life extending options are going to be developed in stages. Perhaps we have already entered the initial stages. Consider the example of knee replacement surgery, which provides the patient with knees that may very well outlast other parts of their bodies, unless eventually the other parts can be replaced too. Perhaps today's increasingly common knee replacement surgery, along with a host of other such routine medical procedures, will one day be viewed as part of a primitive stage in a technological path that led to 500 or more years of healthy knees and indefinitely healthy aging. There is an old saying that death and taxes are two things we cannot avoid. In the end, taxes may prove harder to terminate than death.

Religious Issues

Practical Immortality

"Millions long for immortality who don't know what to do with themselves on a rainy Sunday afternoon."[16] This humorous quip raises a very good psychological question—would humans be overly bored or unsatisfied with centuries of healthy life? The salient truth expressed here is that the will to live, the survival instinct, runs deep in the human psyche.[17]

[15] Ray Kurzweil and Terry Grossman, *Fantastic Voyage: Live Long Enough to Live Forever: The Science Behind Radical Life Extension* (Emmaus, PA: Rodale, 2004). Grossman is a medical doctor. In this book, Kurzweil describes his personal health program, which, e.g., consists of taking about 250 nutritionals a day.

[16] Susan Ertz, *Anger in the Sky* (New York: Literary Classics, 1943), 134.

[17] For discussion of this point, see Calvin Mercer, "Bodies and Persons: Theological Reflections on Transhumanism," *Dialog* 54, no. 1 (March 2015): 30–31; and Noreen Herzfeld, "Must We

Life extension and practical immortality have long been at the heart of the transhumanist vision, and intense passion animates that vision. Nick Bostrom is an important leader in the transhumanist movement. Bostrom, a philosopher who directs the Future of Humanity Institute at Oxford University, said, "Searching for a cure for aging is not just a nice thing that we should perhaps one day get around to. It is an urgent, screaming moral imperative."[18]

In earlier years of the prolongevity movement, enthusiasts sometimes bandied around the term "immortality." Transhumanists eventually found that achieving immortality via technology, rather than through God, generated significant resistance from some faith community members. Prolongevity advocates, who need public support for necessary research, now usually speak in more measured terms. They rarely talk about immortality, leaving that word to religion. Even if people start living hundreds or thousands of years with technological assistance, they can still get hit by a bus, obliterated in a nuclear war, or burned up in a supernova. So, the most we can talk about with technology is "practical immortality." Practical immortality means that one will not die from internal biological causes that have been associated with aging, but one could still die of such things as accidental causes, a new infectious disease, natural disasters, or a cosmic event.

In their effort to garner public support, some transhumanists argue that using technology to achieve lifespans of hundreds of years is not essentially different from what is done now at the local hospital. Medical science has steadily extended the lifespan by eliminating polio, smallpox, measles, and a host of other diseases. The extension of life continues with every treatment advance in cardiac, cancer, and other diseases that take life. "In our efforts to extend life indefinitely, we're just trying to do what's already being done now, but we want to do it more effectively and push the lifespan out much farther," so says the transhumanist.

However, extending life indefinitely (advocated by superlongevity proponents) and extending life by a few years through the compression of morbidity (advocated by current, traditional researchers) have very different implications for the religions. While extending life for a few years does not impact faith claims, rituals, and institutions significantly, radical life extension will most likely evoke significant changes in all these areas. We now consider a few of these possible implications.

Die Transhumanism, Religion, and the Fear of Death," in *Religion and Human Enhancement: Death, Values, and Morality*, eds. Tracy Trothen and Calvin Mercer, 285–99, in *Palgrave Studies in the Future of Humanity and Its Successors*, series eds. Calvin Mercer and Steve Fuller (New York: Palgrave Macmillan, 2017).

[18] "The Fable of the Dragon-Tyrant," *Journal of Medical Ethics* 31, no. 5 (2005), 277. The Future of Humanity Institute was established by the university's largest private donation ever, a gift from James Martin, whose pioneering work led to the "internet of things," addressed in a later chapter.

What Really Matters?

The faith traditions will need to consider how practical immortality might affect spirituality, including experiences of and longing for transcendence, ultimacy, boundlessness, spiritual emotions, and interconnection.[19] If we live for hundreds or thousands of years, will we be as moved by simple wonders of life, like a stunning sunset or an unexpected act of kindness, as we are now? Will we be as inclined to soak up the moment, appreciating and valuing life? Will we continue seeing deeper meaning in places where we might find the sacred? Will we experience, as easily, a sense of transcendence, of something bigger than us that inspires and gives hope, or will we look only to ourselves and our technology for transcendence and deeper meaning? Maybe superlongevity would make us more inclined to experience and appreciate these things since we will have more time. The possibility of radical life extension may push us to become clearer about what we value most and why, or we may become increasingly lackadaisical about values and meaning.

Jewish philosopher and theologian, Elliot N. Dorff, articulates his concern this way:

> …the longer we think we have until we die, the less likely it is that the reality of death will affect our lives. Just as it is very difficult to convince people in their teens and twenties that they need to take their mortality into account, a prolonged life will likely strengthen and lengthen our pursuit of fame and fortune. We will become even more blind to the importance of other values, such as family, enjoying life, fixing the world, and connecting with God.[20]

Turning to the Christian tradition, Protestant and Roman Catholic theologians have expressed concern that living extremely long lives will sap our commitment to and energy for spiritual development. Here's how Protestant theologian, Ron Cole-Turner, puts it:

> Technology offers to give us what we want, or at least what most of us think we want—longer life, youthful bodies, greater health, and mental ability. Christianity invites us to give up what we want, indeed to give up life itself, as the one condition for real life.[21]

[19] For an understanding of mentally healthy spirituality, we draw here upon K. I. Pargament, *Spiritually Integrated Psychotherapy: Understanding and Addressing the Sacred* (New York: Guilford, 2007).

[20] "Becoming Yet More Like God: A Jewish Perspective on Radical Life Extension," in *Religion and the Implications of Radical Life Extension*, eds. Derek Maher and Calvin Mercer (New York: Palgrave Macmillan, 2009; republished in paperback, 2014), 69.

[21] "Extreme Longevity Research: A Progressive Protestant Perspective," in *Religion and the Implications of Radical Life Extension*, eds. Derek Maher and Calvin Mercer (New York: Palgrave Macmillan, 2009; republished in paperback, 2014), 58.

Leon Kass served from 2001–07 as chair of the U.S. President's Council on Bioethics and for years led the bioconservative critique of extreme human enhancement. Along lines similar to Dorff and Cole-Turner, Kass argued that death is a "necessary and desirable end" and gives meaning to life.[22] To remove death, or extend life out indefinitely, would seem to necessitate a significant revisioning of religion, both monotheistic and karmic versions.

How we see death and dying affects our understandings of what makes a technological intervention acceptable and whether radically extending our lives is a good thing. Normative North America is a death-denying culture. Talk of death is avoided, and dying people are often hidden away in hospitals or long-term care homes.[23] Our attitudes and overall worldviews are entwined with our values and help shape our moral reasoning. On the one hand, those who believe there is more to life after death, usually through reincarnation or an afterlife such as heaven, may not be as driven to pursue radical life extending technologies. On the other hand, religions see much that is good about living and will want to extend that, with technology giving more time to do good acts and develop wisdom.

In any case, an argument can certainly be made that different worldviews animate secular technological development and religious commitment to life. That said, it can also be argued that radical life extension through religious salvation and radical life extension through technology need not be mutually exclusive.

Desirability and Acceptability

Turning to the question of the desirability and acceptability of superlongevity, we predict some strange theological bedfellows. Traditionally, liberals and conservatives have lined up in their respective camps on a variety of social, political, economic, and religious issues. With superlongevity, we may see splits among liberals and among conservatives. In other words, bioconservatives and human enhancement technology enthusiasts are likely to show up in both camps.

Liberals embrace science and technology, when it is in the service of good ends, promoting the well-being of people and the planet. Despite their suspicion of science, some conservatives are likely to embrace available therapies and technologies for at least two reasons. First, conservative faith adherents, like humans in general, share a deeply embedded and powerful survival instinct, making the lure of immortality sometimes irresistible. Second,

[22] E.g., "L'Chaim and Its Limits: Why Not Immortality?" *First Things* (May 2001). https://www.firstthings.com/article/2001/05/lchaim-and-its-limits-why-not-immortality.

[23] See, for example, D.P. Waldrop, "Denying and Defying Death: The Culture of Dying in twenty-first Century America," *The Gerontologist*, 51(4), 2011: 571–576. https://doi.org/10.1093/geront/gnr076; C. Adrien, "We live in a death denying culture. That's a problem" [Blog]. 1800 Hospice, 2017. Retrieved from https://www.1800hospice.com/blog/live-death-denying-culture-thats-problem/; and Lucy Bregman, *Beyond Silence and Denial—Death and Dying Reconsidered* (Louisville, KY: Westminster John Knox Press, 1999).

fundamentalists, on the extreme right side of the theological continuum and perhaps especially in Christianity and Islam, can be driven by intense anxiety about going to hell.[24] Living longer on the earth could be seen, perhaps unconsciously, as an opportunity to remain alive until achieving a strong measure of certainty about going to heaven upon death.

Conservatives tend, much more than liberals, to be suspicious of science, particularly when they perceive it as challenging their scriptures or theologies. Evolution is an excellent example of a scientific consensus opposed by some Christian conservatives, especially the more extreme fundamentalists. Since human enhancement technologies emerge out of science, and transhumanist advocates are largely not faith adherents, the general mistrust of science by conservatives may very well bleed over into an opposition, by many conservatives, to the technologies of enhancement. Liberals may oppose radical enhancement on a number of grounds, such as the fair input into, access to, and distribution of the technologies.

Monotheistic Religions[25]

Judaism, Christianity, and Islam are oriented around one life, one death, one afterlife. Living indefinitely will lead to refashioning, in ways that certainly cannot be predicted in detail at this point, of the religions' belief systems, rituals, institutions, and spiritual practices. That said, we speculate that superlongevity will bring considerably more changes to conservative oriented religion than to liberal oriented religion.

As presented in the "Theological Continuum" table,[26] conservative religion, unlike its counterpart on the liberal side, gives attention to, and is at times quite preoccupied with, "otherworldly" matters. Heaven, hell, the final judgement, the soul's fate, and other such topics can outweigh this-worldly concerns. Effectively removing death from the equation would seem to allow focus on matters of this world, which is already of much greater interest to liberal religion.

Even as some conservatives and some liberals embrace much longer lifespans, they will likely make adjustments in scriptural interpretation, doctrines, rituals, and institutions.[27] Addressing death and the afterlife, which is an

[24] See especially Part 3, "A Psychological Profile," (pp. 129–66) in Calvin Mercer, *Slaves to Faith: A Therapist Looks Inside the Fundamentalist Mind* (Westport, CT: Praeger, 2009).

[25] This section is adapted from some of the material in the section entitled "Response to and Impact of Humanity 2.0," in Calvin Mercer, "Insisting on Soma in the Debate about Radical Life Extension: One Protestant's Perspective," in *Transhumanism and the Body: The World Religions Speak*, eds. Calvin Mercer and Derek Maher, in *Palgrave Studies in the Future of Humanity and Its Successors*, series eds. Calvin Mercer and Steve Fuller (New York: Palgrave Macmillan, 2014).

[26] Refer to the "Theological Continuum" table in the chapter, "Transhumanism, the Posthuman, and the Religions."

[27] These elements of religion are addressed by most of the contributors to Mercer and Maher, *Religion and the Implications of Radical Life Extension*.

important part of most religions, especially in their conservative iterations, would be eliminated as a pressing need.

Daily life, for everyone participating in the life-extending therapies and technologies, would be changed in significant ways as well. For example, it may well be that if we live thousands of years, a marriage commitment "until death do us part" may not be as realistic as it is today, given that even today, less than 50 percent of marriages last. Education, careers, birthing children, family life, retirement, and so many other aspects of our lives would also be impacted. Religion all along the theological continuum is integrated into all these aspects in one way or another. So, if religion survives, then extremely long lifespans will necessitate adjustments in the various elements of religion that speak to these aspects of our lives.

The Body

Technologies of radical life extension are based on science that is grounded in a secular, materialistic understanding of the world. The human body can be treated as a machine in this worldview. Religious folk disagree across religions and even within religions about the particular nature of human being but, in general, faith traditions affirm there is more to us than mere flesh and blood. Discussions about what that more is, and how it is understood with respect to flesh and blood, open up interesting, important questions with which we will continue to struggle in the coming chapters.

Keeping the Body and Improving It: The Moderate Scenario

How religions respond to superlongevity is also likely to be informed by the role of the body in life extension. Moderate scenarios of prolongevity entail maintaining the basic structure of our bodies. With de Grey's biological "boring wet approach," for example, we would maintain the same basic physicality as now. Scriptural and theological barriers, in the monotheistic religions, to greatly lengthening life may very well emerge, but they are likely to be less concerning than barriers that would arise with prolongevity visions that fully eliminate the body as we traditionally know it, such as with mind uploading, addressed in a later chapter.

Both liberals and conservatives avail themselves of today's conventional medicines for treating sickness. Rarely do the faithful feel the need to theologically justify normative allopathic medicine.[28] They usually just assume God is working through the hands of medical professionals to bring healing. As the lifespan gradually lengthens, many positive religious views of and use of allopathic medicine will likely continue.

[28] There are a few notable exceptions to this pattern. For example, Jehovah's Witnesses believe that receiving a blood transfusion will damn them to hell, based on interpretations of certain biblical passages (e.g., Genesis 9:4; Leviticus 17:10; Deuteronomy 12:23; Acts 15:28, 29) and their doctrine. If a Jehovah's Witness chooses to receive a donor's blood, they lose their faith community and their relationship with God.

While this is true of conventional medicine, traditional medicines associated with, for example, Indigenous spiritualities, have long been held suspect.[29] Ways of treating illness, including illnesses associated with aging, that are not normative allopathic medicines, have been marginalized, dismissed as useless or even dangerous. Similarly, very new allopathic medical interventions that are breaking new ground, such as genetic modification technologies, may well be regarded as suspect, even if they may eventually become regarded as "normal."

For sure, if medical or technological breakthroughs suddenly make possible life for hundreds of years, with minimal visible changes to our bodies, it may still be difficult for the religions to accept. Scripture and theology in monotheistic religions have been predicated on generally recognizable lifespans, as in the traditional biblical reference to "threescore and ten" (i.e., 70 years).[30] Lifespans of hundreds or thousands of years, especially if the technology came all at once, would perhaps result in the identification of scriptural and theological objections to such technology.

Dropping the Body: The Radical Scenario

A radical scenario for superlongevity would entail significant and visible changes to the physical body or moving into some sort of cyber existence. In the earlier chapter, "Transhumanism, the Posthuman, and the Religions," we considered Max More's "A Letter to Mother Nature: Amendments to the Human Constitution." Amendment 7 of that letter points to the transhumanist interest in enhancements that may take us beyond the bodily existence evolution has bequeathed to us. It reads:

> We recognize your genius in using carbon-based compounds to develop us. Yet we will not limit our physical, intellectual, or emotional capacities by remaining purely biological organisms. While we pursue mastery of our own biochemistry, we will increasingly integrate our advancing technologies into our selves.[31]

In a later chapter, we consider mind uploading, a program yet to materialize, but one which if developed could result in superlongevity in a body/format/platform that merits being called radical. As we will see, this program would entail transferring our memories and personality into some other substrate that is more durable than our bodies. We have been through the ages of stone, bronze, and iron. Silicon may be the preferred material for today's digital world; by the time mind uploading arrives, some other material may have taken silicon's place.

Ray Kurzweil's more general vision is well-known with regard to our future "bodies."

[29] M. Hanrahan and B. Wills, "Makayla's Decision: The Exercise of Indigenous Rights and the Primacy of Allopathic Medicine in Canada," *The Canadian Journal of Native Studies* 35, no. 2 (2015): 207–223.

[30] Psalm 90:10.

[31] http://strategicphilosophy.blogspot.com/2009/05/its-about-ten-years-since-i-wrote.html.

> By the time we have the tools to capture and re-create a human brain with all of its subtleties, we will have plenty of options for twenty-first-century bodies for both nonbiological humans and biological humans who avail themselves of extensions to our intelligence. The human body version 2.0 will include virtual bodies in completely realistic virtual environments, nanotechnology-based physical bodies, and more.[32]

> I envision human body 3.0—in the 2030s and 2040s—as a more fundamental redesign. Rather than reformulating each sub-system, we (both the biological and nonbiological portions of our thinking, working together) will have the opportunity to revamp our bodies based on our experience with version 2.0. . . One attribute I envision for version 3.0 is the ability to change our bodies. We'll be able to do that very easily in virtual-reality environments ... but we will also acquire the means to do this in real reality. We will incorporate MNT-based fabrication [MNT is molecular nanotechnology] into ourselves, so we'll be able to rapidly alter our physical manifestation at will.[33]

In these radical scenarios of life extension, all the norms and institutions of society will undergo major changes, and religion, if it survives, will not be exempted from changes as well. Such radical visions would unfold in stages, if at all, giving the religions time to try to adjust. However, we should not minimize the measure of this kind of disruption. We are envisioning an evolution that would involve separating who we are from that aspect of ourselves with which we are quite intimate, namely, our bodies.[34] The resistance to such a development, and its impact on religion, should not be underestimated. The religions, at the very least, are likely to urge great caution if these radical scenarios become feasible.

From a monotheistic perspective, the more specific issue with regard to scenarios that eliminate or radically alter the human body has to do with the scriptural and theological importance attached to physicality, both as a constituent part of who we are as God's creation and as a part of God's activity in the world. To give Christianity as an example, materiality is a meaningful part of the stories of creation, incarnation, resurrection (i.e., of the body), and the end-time (e.g., a new heaven and a new earth). The theological anthropology of these religions is one of embodiment in a human body that we have known since God created it from the dust.

Of course, these monotheistic religions assert that the human body is not mere material, unintegrated with the rest of the person. The body, it is believed, is enlivened by God's breath, God's spirit. But, rejecting materialism does not

[32] Ray Kurzweil, *The Singularity Is Near: When Humans Transcend Biology* (New York: Viking, 2005), 199.

[33] Ibid., 310.

[34] Phrases like "who we are" and "that part of ourselves" raise complicated and important questions about human nature and personal identity. We will address these in some detail in the later chapter on mind uploading.

mean the religions can dispense with the human body without some fundamental shifts in scriptural interpretation and theology.

Having made the case that monotheistic religions, in the main, affirm the body as important to who we are, we now turn to one stream of the Christian tradition that perhaps might be well positioned to embrace, at least with regard to physicality, a radical scenario where the body is eliminated in favor of some sort of cyber existence. We provided the background to this view in the chapter, "Transhumanism, the Posthuman, and the Religions."

As discussed in that earlier chapter, we saw that most Christians do not believe it is possible or even desirable to separate the body from the rest of the person. The doctrine of the incarnation includes the conviction that Jesus was fully divine and fully human. For most Christians, the incarnation is evidence that people are divinely created, and irrevocably integrated. The whole person dies at the end of their life and is subsequently resurrected fully in a new and transformed body.

However, a minority Christian tradition, exampled by many conservative Protestants today, is influenced by a dualistic outlook that sees the body as separate from the soul. The body can die, but death does not affect the essence of the person, the soul, which lives on basically unchanged. This dualistic understanding (i.e., the human is some combination of two otherwise disparate parts, body and soul) is actually quite compatible with mind uploading, which attempts to capture the essence of the person in the download that can then exist indefinitely. The conservative reluctance toward science is still a barrier, but it will be interesting to see if the mind uploading path to superlongevity is given any credence by this conservative wing of the religion.

Finally, both conservatives and liberals, albeit in their own ways, affirm the importance of community. We are social creatures. If we detach our minds or souls from our bodies, perhaps that will negatively impact community life and relationships. Or, maybe a disembodied community opens up higher levels of community. Certainly, the institutions that support community and the rituals that animate it will change.

Karmic Religions

Monotheistic and karmic religions articulate strikingly different notions about the afterlife. We addressed, in general, these different orientations in the chapter titled "Transhumanism, the Posthuman, and the Religions." We suggest you briefly review the ideas of karma and reincarnation explained in that earlier chapter. Now, we drill down into the particulars of reincarnation and its implications for extreme longevity.

There is reason to think that extreme longevity may have less impact on the karmic religions than on the monotheistic ones. First of all, death is already of less importance in the karmic religions since, as one Buddhist scholar puts it, "... the death in one particular lifetime is encumbered with less gravity; after all, another lifetime is just around the corner."[35] Buddhism's doctrine of no-self (Sanskrit, *anatman*) means this religion is perhaps even less concerned about mortality than are the other karmic religions. Death certainly gets attention in the theology, rituals, and psychology of karmic religions, but we can anticipate that removing it will have less impact upon beliefs than in the monotheistic traditions.

Whereas the law of karma (i.e., one's actions determine the next rebirth) is almost always applied to future lives, radical life extension might prompt theologians of the karmic religions to give more attention to the impact actions can have later on in the current lifetime, since that one lifetime will extend indefinitely. The cause and effect aspect of karma is strengthened in the immediacy of one's life, while the rebirth component fades from view, at least for all practical purposes, since life continues indefinitely.[36]

Finally, turning to one's personal spiritual life, karmic religion scholars, as we have noted, have not yet addressed superlongevity to the same extent as have theologians and religion scholars of monotheistic religions. However, those scholars who have addressed the karmic religions and radical life extension share a concern that superlongevity could negatively impact spiritual motivation and aspiration. Buddhist scholar Derek Maher writes:

> Buddhists would need to elaborate new rhetorical strategies to encourage people to engage in religious practice. If people could expect to live forever, they would have fewer incentives to behave well and foster good karma since they would no longer fear being reborn as headlice, dung beetles, or worse, hell beings.[37]

Ted Peters is a major and widely respected voice in the theological and ethical assessment of radical enhancement. After reviewing contributions from scholars writing about karmic religions, Peters concludes that "although the fundamental worldview will remain intact, we can predict a deterioration in motivation for pursuing the more sublime spiritual goals."[38] Of course, one can argue that a longer life provides opportunity for more time to deepen the formation of spiritual life. The initial reflections by scholars writing about karmic and monotheistic religions on superlongevity, however, are cautious and pessimistic about spiritual well-being in the face of extremely long lives.

[35] Derek Maher, "Two Wings of a Bird," in *Religion and the Implications of Radical Life Extension,* eds. Derek Maher and Calvin Mercer (New York: Palgrave Macmillan, 2009; republished in paperback, 2014), 114.

[36] Derek Maher, "Two Wings of a Bird," 120; and Arvind Sharma, "'May You Live Long:' Religious Implications of Extreme Longevity in Hinduism," in eds. Calvin Mercer and Derek Maher, *Religion and the Implications of Radical Life Extension* (New York: Palgrave Macmillan, 2009; republished in paperback, 2014), 151–52.

[37] Maher, "Two Wings of a Bird," 119.

[38] Ted Peters, "Reflections on Radical Life Extension," in *Religion and the Implications of Radical Life Extension,* eds. Derek Maher and Calvin Mercer (New York: Palgrave Macmillan, 2009; republished in paperback, 2014), 163.

Ethical Issues

Discussions about superlongevity often move directly to objections. Perhaps we are stating the obvious, but we begin by briefly putting on the table that there are obvious benefits to longer, healthy living. Such benefits include decreased suffering, more opportunities to enjoy the benefits of a healthy life, and giving people choice and power over their life and death.[39] In the earlier chapter, "Radical Human Enhancement and Ethics," we discussed the difference between precautionary and proactionary approaches to superlongevity research. These stances yield different results in weighing the costs and benefits.

Those advocating a primarily precautionary approach highlight the concerns discussed later in this section on ethics, as well as unknown damaging side effects of the research programs. Proactionary advocates insist the benefits of human prolongevity development outweigh the risks. We discussed at some length the role of values in decision-making in the earlier chapter on ethics. Keep all these important ethical considerations in mind as we summarize some of the more common objections, and possible responses, to superlongevity programs.

The Therapy—Enhancement Continuum: What It Means to Make Us Better

When radical life extension is framed primarily as a therapy—enhancement issue, questions about a technology's fit with religious integrity emerge. Consider the hypothetical example of an approved and regulated organ replacement with 3-D technology.[40] Currently, many people die daily waiting for a suitable organ. No doubt we would consider 3-D generated organs for those in need of transplants as therapeutic and, therefore, ethically permissible, at a minimum. On the other hand, as we age our organs can become less efficient and damaged. Perhaps it would move beyond therapeutic, to enhancement, to use 3-D generated organs for all people in later life, if they choose them.

The religions would need to consider if life-extension for elderly adults, via 3-D technology, is congruent with faith claims and values. It is possible for a technology to land clearly in the enhancement side of the continuum and still be considered religiously congruent and, therefore, acceptable. But harder questions will likely need to be asked of technology that lands in this end of the continuum, since technology in the enhancement end of the continuum is less clearly about healing depending, of course, on how one defines aging.

[39] An excellent summary of the "case against death" is the following book by philosopher Ingemar Patrick Linden, *The Case Against Death* (MIT Press, forthcoming).

[40] Emma Yasinski, "On the Road to 3-D Printed Organs," *The Scientist—Exploring Life, Inspiring Innovation* (February 26, 2020). https://www.the-scientist.com/news-opinion/on-the-road-to-3-d-printed-organs-67187. Artificially generated organs would presumably not present the immune-rejection issues that are part of using a donor organ, no matter how well matched.

Our conclusions about a particular intervention rests, to a large degree, on how we interpret aging. If aging is interpreted as a disease, then the elderly would be seen in need of healing. Currently, aging is not seen by the general population and the medical establishment as a disease. Transhumanists work hard promoting the idea that aging is a disease that should be eradicated.

Disability theologians help us raise additional questions about who decides, and how it is decided, what constitutes a disease. Some argue editing out trisomy 21 (Down's Syndrome) is therapeutic. Others contend that by editing out Down's Syndrome we are reinforcing prejudice, and amplifying patterns of injustice, against those with abilities different from the norm. We may be failing to value people with different disabilities if we choose to make people "better" by eradicating these perceived disabilities or diseases. If interventions are judged to be unjust, then they cannot be considered therapeutic, in the fullest sense of that word.

While there is plenty of debate about particular interventions, religion is pro-healing. So, any intervention that is considered clearly therapeutic is likely to be supported by religion, all else being equal. How we decide, and who decides, what is therapeutic will continue to be problematic. Most interventions clearly falling on the enhancement side of the continuum, however, will elicit considerable disagreement. What is embraced by religion is a moving target, because the range of healing options has greatly increased over the centuries. We now accept many previously unthinkable interventions, such as heart surgery, as normal and therapeutic. In other words, it is common for enhancements to become perceived as healing therapies over time, if potential harms are outweighed by potential benefits.

Key to a sound moral assessment of interventions from a religiously-based perspective is how the interventions impact relationships and, indeed, the well-being of all life. The monotheistic religions believe that God can work through people to create for the good of self, neighbor, and creation. In an earlier chapter, we introduced the concept of "created co-creator," coined by theologian Philip Hefner.[41] Some theologians in the Judeo-Christian tradition have turned to this concept to explain their acceptance of life extension programs.

Disability theologians make the case for human fragility and vulnerability as virtues that compel us into relationship.[42] The case for superlongevity is stronger if it is supportive of the deepening and expansion of authentic relationships. If we overcome physical fragility and vulnerability, it may be that we will need to pay more attention to the other ways in which we are fragile and vulnerable in order to enhance relationships. Relationships are important in religions. Becoming better humans will mean attending to the possible negative relational implications of overcoming physical vulnerability.

[41] Philip Hefner, *The Human Factor: Evolution, Culture, and Religion* (Minneapolis, MN: Fortress, 1993), 27.

[42] E.g., Thomas E. Reynolds, *Vulnerable Communion: A Theology of Disability and Hospitality* (Grand Rapids, Michigan: Brazos Press, 2008).

Choice

Anti-aging proponents highly value morphological freedom and often frame radical life extension as an issue, primarily, of individual choice.[43] If we decide we want the pill or any enhancing technology that extends our lives, and we can pay for it, we should be able to choose to have that technology. Prolongevity leaders do recognize that not everyone can financially afford these enhancements, and they accept that as a temporary limitation.

The issue of free choice raises sharply the distinction between an individual choosing an enhancement for themselves and making these choices on behalf of progeny. This distinction finds particular expression in potential germline genetic modification technologies. Germline gene modification is extremely controversial, since this technology targets the genes in germline cells, including sperm, eggs, and embryos, which are passed from generation to generation.

Germline genetic modification that prevents diseases caused by genetic mutations, such as muscular dystrophy, Down's Syndrome, and cystic fibrosis, are considered to be potentially more acceptable than modifications that would radically enhance progeny. However, even disease prevention is complicated by unintended off-target effects of the gene therapy, yielding risks that would be assumed by all progeny.

Regulatory agencies have generally drawn the line at germline genetic modification in part because of consent issues and also because of possible negative side-effects. As of 2021, germline genetic modification was prohibited in 40 countries. Conceivably, somatic genetic modification technologies could proliferate and become perfected to the point that scientists conclude these therapies pose little risk if used in germline programs for progeny. While germline therapies to prevent deadly diseases seem desirable, the water turns murky when considering germline therapies preventing genetic conditions that may be uncomfortable but not deadly. The slippery slope concern applies here. Perhaps parents who dream of their child excelling at athletics could use stem cell therapy to enhance muscular strength. But it may be that the child has no interest in athletics or muscular strength. Another legitimate concern is the possibility of unintended and unanticipated off-target effects, some of which may not actualize for a generation or more.

There are other complications regarding consent, as an expression of autonomy, in addition to the challenges posed by heritable enhancements. As we discussed in the chapter on ethics, complex social processes are at work influencing our identities and choices. How do we know what we really want? To illustrate, a normative North American understanding of "successful aging" is that aging adults continue to make productive contributions to society and to

[43] Anders Sandberg, "Morphological Freedom—Why We Not Just Want It, but Need It," in *The Transhumanist Reader: Classical and Contemporary Essays on the Science, Technology, and Philosophy of the Human Future*, 56–64, eds. More, Max and Natasha Vita-More (West Sussex, UK: Wiley-Blackwell, 2013).

function independently.[44] Since technology is embedded with the value of utility, its use adds to this normative view about aging. Utility can be valuable, but religions provide a balance, affirming the intrinsic worth and dignity in people, regardless of their ability to produce something measurable.

Religions also celebrate our interdependence, contending that dependence can be seen as a virtue rather than something to overcome. Surrendering one's life to God is, generally speaking, encouraged in all religions, monotheistic and karmic. We are dependent on God and, in similar fashion, we live in community, dependent and supportive of one another.

These beliefs about the dignity and worth of everyone and interdependence do not necessarily mean religious people will not choose radical life extending technologies. The beliefs, however, do present parameters for making enhancement choices. In ethics language, autonomy is understood as relational; my choice to live longer-or-not is not just about me. This emphasis on relationality shifts the conversation from rights to responsibilities. So, instead of asking what my rights are, I need to ask what my responsibilities to others and the world are.

Justice

As explained in the chapter, "Radical Human Enhancement and Ethics," the world's religions are generally united, at least in their scriptural and theological traditions, in their ethical objection to social disparity along race and class lines. That is a general statement that must be applied and nuanced in each religious situation.

For example, Hinduism's doctrines of karma and reincarnation have been critiqued for what could be interpreted as theological justification for social disparity exhibited in the traditional caste system. That critique has to be balanced by appreciating the efforts of social reformers, such as Mahatma Gandhi, who in speech and practice strenuously opposed caste. Christianity has an unsavory history of oppressing people under the twin colonial banners of sword and cross. Yet, the justice teachings of the biblical prophets and Christ inspired Martin Luther King, Jr., an important reformer in the United States civil rights movement who also visited and learned nonviolence civil disobedience tactics from Gandhi.

Superlongevity could conceivably exacerbate social disparity. Using economics to illustrate, much of the world's wealth is in the hands of older people. People of means are the very ones who would have access to prolongevity interventions, likely to be expensive. Radical life extension, then, could very well further concentrate the world's wealth in the hands of the rich, since they would be among the first to live indefinitely. Given the positive correlation between race and class, superlongevity in this scenario yields greater social disparity in many categories.

[44] Martha Holstein, J. Parks, and M. Waymack, *Ethics, Aging, & Society: The Critical Turn* (New York: Springer Publishing Company, 2011), 45–64.

The above noted concern about social disparity easily leads to distributive justice issues. As we have explained, scholars and theologians of the karmic religions have not yet weighed in strongly on radical human enhancement technology. However, we can anticipate that when they do, one of the early concerns is likely to be about the fair distribution of these powerful technologies. We can see that expressed by Buddhist scholar Derek Maher.

> For Buddhists, a variety of ethical considerations attend the application of the biomedical innovations under discussion. The most notable concern would be that such advances ought to be made available on an equitable basis. Most Buddhists would condemn a program through which only the wealthy, powerful, or well connected were able to take advantage of prolongation therapies.[45]

Concerns about distributive justice are particularly prominent among people of faith who tend toward the liberal or progressive wing of the religions, and these questions regarding equitable access may actually be their main objection. We saw in the chapter, "Transhumanism, the Posthuman, and the Religions," that advocacy for the disenfranchised runs deep in the Abrahamic faiths, especially as interpreted by liberals. Terminating aging is likely to be an expensive project. Liberals especially will worry about compounding the problem of distributive justice by the diversion of resources from basic needs, such as clean water, health care, education, and other programs that help poorer communities.

One response transhumanists often make to this critique is that superlongevity and radical enhancements are being singled out unfairly. As put sharply on one transhumanist website,

> Those asking such questions should begin by giving up *their* late model car, *their* cable TV, and *their* cappuccino for the benefit of the underprivileged of the world before asking others to give up their life![46]

A second response to distributive and social justice objections is that the superlongevity option is not going to arrive overnight. A more likely scenario is something like the following. A story will hit the front pages of newspapers that scientists have quadrupled the lifespan of a mouse. Because of the transferability of research from mice to humans, people will then realize that, yes, extreme longevity can happen for humans. It could take a decade or more to transfer that breakthrough in mouse lifespan to humans. During that time, people all over the world will demand access to this technology when it arrives. No government or wealthy class will be able to hold the technology private. The survival instinct is just too strong.

[45] Maher, "Two Wings of a Bird," 120. From a Western religion, Roman Catholic scholar Terence L. Nichols expresses the same concern in "Radical Life Extension: Implications for Roman Catholicism," in *Religion and the Implications of Radical Life Extension,* eds. Derek Maher and Calvin Mercer (New York: Palgrave Macmillan, 2009, republished in paperback, 2014), 140–44.

[46] Alcor Life Extension, "Frequently Asked Questions." https://alcor.org/FAQs/faq03.html#world.

Counterarguments to this response include the collective will argument that food is necessary for survival, and the capability exists to produce enough food for everyone, but people still starve, due to political and economic selfishness. We have the means but lack the collective will to move significantly on a host of social justice issues. Superlongevity will not be any different, the critic contends. Another counterargument is that any length of trickle-down time is unacceptable. The rich will get richer and more powerful with extended lives. The gap between the rich and poor or not-so-rich will widen during the trickle-down period.

A New Division in a Socially Unjust World

Another major concern about superlongevity and other enhancements is that radical enhancements will create yet another category of division in the human population. When some humans are radically enhanced and not others, it will lead to a division in society between the enhanced and the unenhanced, or "normals."

Possible long-range implications of some people living hundreds of years while others live "four-score and ten" are troubling. We already noted earlier in this chapter that wealth, acquired in the first hundred years of life, could aggregate and multiply over thousands of years, leading to even more disparity in wealth. Moreover, countries where the technology companies are located are likely to acquire more political power globally.

While concerns about superlongevity aggravating wealth and power disparities are heightened, especially among liberals in religion, Donna Haraway and others have argued that we will change our appearance through technology as we become cyborgs. These changes may in a helpful way blur the divides not only between human and machine, but also between different skin colors, genders, body size, and other visible characteristics that have been used to unjustly discriminate and judge. A cyborg is described by Haraway as "a hybrid of machine and organism, a creature of social reality as well as a creature of fiction."[47] Haraway sees the cyborg as "a creature in a postgender world," transgressing the humanly constructed categories, in particular those of human and machine that we now use to understand each other.[48] While some have seen Haraway's work, and in particular her 1985 "A Cyborg Manifesto,"[49] as necessarily supportive of transhumanism, Haraway does not agree.[50]

[47] *Simians, Cyborgs and Women: The Reinvention of Nature* (New York: Routledge, 1991), 4.

[48] Ibid., 8. See also Aleksandra Lukaszewicz Alcaraz, *Are Cyborgs Persons? An Account of Futurist Ethics*, in *Palgrave Studies in the Future of Humanity and Its Successors*, Calvin Mercer and Steve Fuller, series co-editors (New York: Palgrave Macmillan, 2021); and J. Jeanine Thweatt-Bates, "Artificial Wombs and Cyborg Births," in *Transhumanism and Transcendence: Christian Hope in an Age of Technological Enhancement*, ed. Ron Cole-Turner (Washington, DC: Georgetown University, 2011), 105–108.

[49] Republished in Haraway, *Simians, Cyborgs and Women*.

[50] For a helpful commentary on Haraway and transhumanism, see Steve Fuller, *Nietzschean Meditations: Untimely Thoughts at the Dawn of the Transhuman Era*. Posthuman Studies 1, ed. Stefan Lorenz Sorgner (Basel: Schwabe Verlag, 2019), 112–18.

Re-embodiment that will come with some radical enhancements has liberatory potential if—a hugely important "if"—values and attitudes that give rise to prejudice in the first place are intentionally and successfully transformed. Accomplishing this will not be easy. We see that search engines and AI algorithms mirror and reinforce racism and sexism.[51] Likewise, digital "body" platforms that erase flesh, blood, and skin will not necessarily erase prejudice and stereotypes.

Ageism

Many people experience being discounted, not heard, talked down to, or otherwise devalued as they develop grey hair, wrinkles, limited mobility, and other visible signs of aging. Ageism is a serious social justice issue. A glaring example of how attractive technology can be portrayed and how unattractive aging can be portrayed, is the Kia commercial unveiled during a Super Bowl. As Aerosmith's 1973 hit "Dream On" plays in the background, lead singer 69-year-old Steven Tyler races a new Kia Stinger and becomes young again. Kia's tagline in the commercial was "Feel something again." The ad suggests that as we age we lose vibrancy, becoming irrelevant and boring, unable to be stimulated and engaged. Superlongevity might inadvertently magnify ageism, feeding into an assumption that older adults are dull, out of touch, and in need of technological fixing.

The obvious transhumanist response to this concern is that aging does not have to be experienced in frail, diseased-ridden bodies that suffer and die. The transhumanist vision is that prejudice against the elderly will actually be eliminated when the elderly are living healthy, vibrant, active lives.

Longevity Dividend

Most of the ethical discussion about superlongevity thus far has been objections or concerns. The "Longevity Dividend," however, is the idea that there can be positive social justice and societal well-being benefits from superlongevity because of the financial savings. The term was coined by S. Jay Olshansky and colleagues.[52] Olshansky is a public health scholar with specialization in gerontology and biodemography. Biodemography is a new interdisciplinary field that addresses biological and demographic factors that impact birth and death in shaping individuals and populations.

This social justice argument for superlongevity is very old, illustrated by a 2006 initiative explaining the concept and its social, economic, and political dimensions. In that year, at an event on Capitol Hill, United States senators from both sides of the aisle, Nobel Laureates, representatives of national and

[51] Trishan Panch, H. Mattie, and R. Atun, "Artificial Intelligence and Algorithmic Bias: Implications for Health Systems," *Journal of Global Health* 9, no. 2 (2019): 010318. https://doi.org/10.7189/jogh.09.020318.

[52] S. Jay Olshansky, "Reinventing Aging: An Update on the Longevity Dividend," *Aging Today: A Bimonthly Newspaper of the American Society on Aging* (March/April 2013). https://www.asaging.org/blog/reinventing-aging-update-longevity-dividend.

international health organizations, and scientists delivered to representatives of Congress a petition on behalf of the Longevity Dividend. The advocates attempted to convince elected officials that significant public funds could be saved if people live healthy longer, because most medical dollars are usually consumed in the last years of life, during failing health. The call to action for public officials is to invest in "anti-aging research," as opposed to mere "disease research."

Population Explosion

An ethical concern often expressed is some form of, "Since we already have more people than the planet can support, we're not going to be able to handle people living for hundreds of years."[53] A number of responses to this concern have been offered; we make no judgment about the adequacy of these solutions.

Theoretically, a law could mandate that anyone partaking of longevity technology is prohibited from having children. A legal solution would not be as simple as making a law, of course, but there could be public policy initiatives that to some degree mitigate the overcrowding problem. We know that limiting the number of children a couple may have has been instituted by some countries (e.g., China) for various reasons.

The solution of eliminating future generations raises its own ethical issues. Do we have a responsibility to allow future generations to come into being at all? Some transhumanists think it is absurd to require fully functioning adults, or those who with technology could be fully functioning, to move off the stage of life to make way for those yet unborn. Steve Fuller presents the case for "generational change as a vehicle for radical conceptual change," providing for "periodic rejuvenation."[54]

> A younger generation is supposed to be more open-minded to new experience and hence more easily impressed by whatever happens, precisely because youth sets fewer prior expectations about what should happen. In sociological terms, the young are not so heavily invested in the past, even if they have been formally trained in established forms of knowledge, because they have not spent so much time routinely enacting their training as part of their self-presentation—and hence have not personally experienced the benefits of sticking to the old ways.[55] … Contemporary transhumanism, despite its much vaunted 'futurist' vision, remains largely oblivious to the intergenerational consequences of its quest for indefinite longevity.[56] … the mind-set of the younger generation–the sense of exhilaration

[53] Muslim scholar Aisha Y. Musa asks that question. See "A Thousand Years, Less Fifty: Toward a Quranic View of Extreme Longevity," in ***Religion and the Implications of Radical Life Extension***, eds. Derek Maher and Calvin Mercer (New York: Palgrave Macmillan, 2009. Republished in paperback, 2014), 129.

[54] Fuller's discussion of this idea is in *Nietzschean Meditations*, 175–86.

[55] Ibid., 177.

[56] Ibid., 184.

that comes from acting in a bold yet naïve manner–may be threatened with extinction.[57]

A second solution offered is colonization of outer space. At first blush, this may seem like a fanciful solution, more tailored for a science fiction movie. However, one common problem in objections to radical human enhancement is that detractors engage the conversation as if technological development remains static, except for the specific technology being addressed. A technological advance in one area is likely to be matched over time by advances in other areas. Space travel is expected to become increasingly common. While a long way off, if ever feasible, space colonization is proposed as a solution to overpopulation. Should it ever materialize on a large scale, settlements in space could address overpopulation, as well as other natural or human-made planetary disasters.[58]

The concern about population, when combined with the other justice concerns of ageism and a new division between the normal and the enhanced, provides support for a precautionary stance toward developing superlongevity technologies. Proactionary advocates, however, as noted, minimize these dangers while pointing to the significant longevity monetary dividends, as well as providing people with less suffering and greater choice over their life and death. As we have emphasized, the values brought to the process help shape the analysis.

Questions for Discussion

1. If people begin living for hundreds of years, how might this impact rituals and institutions in religions with which you are familiar? For example, will we still be expected to marry someone for life? How might end of life rituals change?
2. With extreme longevity, how might death be reconceptualized and addressed in various religious rituals and in the personal experience of faith adherents? How might religious leaders revise their guidance of members of their communities?
3. Discuss the meaning of aging. Is aging a disease? Is it a desirable part of being human?
4. Do you think extreme longevity will affect the notions of reincarnation or heaven in the different religious traditions? Why or why not?

[57] Ibid., 186.

[58] For an exploration of this option by an expert, theoretical physicist Michio Kaku, see *The Future of the Mind: The Scientific Quest to Understand, Enhance, and Empower the Mind* (New York: Doubleday, 2014). The book's three sections are titled "Leaving the Earth," "Voyage to the Stars," and "Life in the Universe."

5. To give focus, we have concentrated in this chapter on the physical enhancement of longevity. Longevity, while the most talked about, is only one of many possible physical enhancements. What others might be possible, and what might be some of their religious and ethical implications?
6. What theological and ethical concerns, in addition to the ones we discussed, can you imagine with regard to radical physical enhancement? Consider the responsibilities that would come with living radically extended lives. What might these responsibilities entail?
7. Do you think a stronger case can be made for a precautionary or proactionary stance with regard to radical human enhancement? Why? What values played a role in your answer?
8. Would you take the longevity pill? Why or why not? Is your answer informed by a religious view? If so, what?

CHAPTER 6

Cognitive Enhancement and Moral Bioenhancement: Becoming Smarter and More Moral

Technology of Cognitive Enhancement

Perhaps you remember taking the Scholastic Aptitude Test (SAT), a test necessary for college application in the United States, or scrambling to pump up your Grade Point Average (GPA). Would it not have been great to have a cognitive enhancement pill that could add a couple of hundred points to your SAT score or boost your GPA? Maybe yes, maybe no.

As we proceed, we place radical cognitive enhancement in the larger context of traditional methods for cognitive enhancement. Extreme cognitive enhancement, together with artificial intelligence (AI), may necessitate moral enhancement to ensure that we use our increased intellect for good. As we increasingly overcome what have been considered natural cognitive limits, it is imperative to consider the theological and ethical implications.

How We Make Ourselves Smarter

Cognitive enhancement is not new. We work hard in many ways to make ourselves and our children smarter. Mostly, we promote behavioral methods for cognitive enhancement, such as good study skills and education. Long standing education techniques are continuing to be improved and utilized. Mental or psychological training, such as that provided by guided imagery, mindfulness exercises, and mnemonics and other memory techniques, have long been used to improve our mental abilities. Learning another language, recreation, eating well, appropriate sleep, and sports psychology techniques also contribute. However, we have not been satisfied with these behavioral techniques, perhaps because they require time and hard work. We want more. And we want more for less effort.

Leading futurists Nick Bostrom, director of the Oxford University Future of Humanity Institute, and his colleague, Anders Sandberg, define cognitive

C. Mercer, T. J. Trothen, *Religion and the Technological Future*,
https://doi.org/10.1007/978-3-030-62359-3_6

enhancement as using artificial means to optimize learning and memory systems.[1] Many cognitive enhancers also augment our bodies and, as we see in a later section, even our moral selves. In an information society, increasing focus is put on cognitive capacity and information access and analysis.

Experts point out that cognitive enhancement is complex, with several factors needing to be considered.[2] As we will see, there are many methods of cognitive enhancement. Also, cognition is not one thing. It is a complex assemblage of capabilities, including attention, memory, processing speed, creativity, and cognitive flexibility. Complicating things further is the role of social acceptableness. Consider, for example, the use of pharmaceutical drugs that people procure in ways other than via a prescription from their doctor.

Despite the complexities and complications, people have always attempted to become smarter and that attempt is now needing assessment, religious and otherwise, as powerful options for cognitive enhancement are being developed. Before too quickly judging radical cognitive enhancement, consider the wide use of night-long cognitive enhancements prior to big university exams. Street names for these cognitive stimulants, long used, are coffee and cola. Jolt, a caffeine and sugar intense cola, became available in 1985, marketed to students and young professionals with the slogan, "All the sugar, twice the caffeine!" Caffeine and sugar (and Jolt equivalents such as Red Bull) are still very much in use, complemented by today's array of technological aids like smart phones, calculators, and internet, all of which extend our cognitive reach.

As with many enhancements we will discuss, genetic engineering will almost assuredly play a role in cognitive enhancement. As it becomes more common to screen embryos for genetic markers of disease, some parents are going to want to screen for markers associated with IQ. Manipulating the genes of mice has already greatly improved their cognitive abilities. The transferability of research from mice to humans will put genetics center stage in the cognitive enhancement debate. Reducing cognitive decline due to aging may be viewed favorably as therapeutic, but drastic enhancements in IQ, beyond what is considered normal, will be controversial.[3] In addition to genetics, we now turn to a number of other interventions to make us smarter.

Smart Drugs and Nootropics

Methylphenidates, like Adderall and Ritalin, are stimulants that sharpen cognitive focus and facilitate alertness, memory, and wakefulness, although is it

[1] Nick Bostrom and Anders Sandberg, "Cognitive Enhancement: Methods, Ethics, Regulatory Challenges," *Science and Engineering Ethics* 15 (2009): 311–41. https://doi.org/10.1007/s11948-009-9142-5.

[2] M. Dressler, et al., "Hacking the Brain: Dimensions of Cognitive Enhancement," *ACS Chemical Neuroscience* 10, no.3 (March 20, 2019):1137–1148.

[3] "Should We Pursue Genetic Cognitive Enhancement?" *The Hastings Center News* (April 3, 2018). https://www.thehastingscenter.org/news/enhanced-human-risks-opportunities/.

questionable whether these stimulants actually improve academic performance.[4] Methylphenidates help restore an imbalance of neurotransmitters in people with attention deficit hyperactivity dissorder (ADHD). Because these drugs are misused by people without ADHD, they often end up backfiring and actually hamper good thinking after the negative and positive effects are taken into account. Evidence so far indicates that methylphenidates do not improve academic performance. Motivations for taking these "study drugs" include coping with low self-esteem, a need for external validation, having few sources of meaning in life, insufficient parental attention, and mental health issues, including psychosis.[5]

Stimulants also improve reflexes and blunt appetite, which can be (dangerously) attractive to people who wish to lose weight. A recent study found that more than five percent of North American university students use ADHD treatment drugs, such as Adderall or Ritalin, to help them study, in spite of potential cardiac problems, addiction, and increased anxiety. The use of study drugs for non-medical reasons is growing at Canadian and United States universities.[6]

Other pharmaceuticals sometimes used to facilitate faster and better thinking include modafinil (a common trade name is Provigil), which is often prescribed to people with sleep disorders such as narcolepsy. What is sometimes missed by off-label and self-prescribing users are the potential harmful side effects, including headaches and insomnia. Most people can take modafinil only for a very limited time before the risks outweigh the benefits. Antidementia drugs, such as acetylcholinesterase inhibitors and memantine, are also theorized to have cognitive benefits for people without dementia, but that is controversial and scientifically unproven. The value some people place on any means to cognitively enhance is of serious concern to medical professionals, since self-prescribed usage of these drugs can be very dangerous.

Nootropics (Greek for "mind turning") are widely used smart drug supplements.[7] "Natural" cognitive enhancement is a big industry, with internet advertisements for a host of supplements, such as omega-3 fatty acids, ginkgo biloba, Lumalta, and the supposed brain booster pill Geniux. Claims that Geniux can significantly boost cognitive enhancement have been debunked,

[4] Dressler et al., "Hacking the Brain."

[5] D. D. Abelman, "Mitigating Risks of Students Use of Study Drugs Through Understanding Motivations for Use and Applying Harm Reduction Theory: A Literature Review," *Harm Reduction Journal* 14, no. 68 (2017). https://doi.org/10.1186/s12954-017-0194-6.

[6] K. R. Holloway, T. H. Bennett, O. Parry, and C. Gorden, "Misuse of Prescription Drugs on University Campuses: Options for Prevention," *International Review of Law, Computers & Technology* (2013) 27:324–34. https://doi.org/10.1080/13600869.2013.796707; and C. R. Skidmore, E. A. Kaufman, and S. E. Crowell, "Substance Use Among College Students," *Child and Adolescent Psychiatric Clinics in North America* 25 (2016): 735–53. https://doi.org/10.1016/j.chc.2016.06.004.

[7] Nayef Al-Rodhan, "The Runaway Train of Cognitive Enhancement," *Scientific American*, (December 9, 2019). https://blogs.scientificamerican.com/observations/the-runaway-train-of-cognitive-enhancement/.

with refund cheques issued by the United States Federal Trade Commission.[8] Despite the debunking of Geniux, numerous other supplements, with similar claims about improved focus, memory and energy, are being promoted, often without scientific validation.

Brain Biohacking

Neuroscience developments in brain stimulation techniques have generated much interest among athletes and others who want to overcome mental messages that inhibit physical performance. For example, transcranial direct current stimulation (tDCS) improves endurance by making it easier to overcome mental messages regarding pain or exhaustion that would otherwise encourage the athlete to stop. tDCS is rumored to have been used in past Olympic Games.[9] Harms are associated with tDCS, including seizures, headaches, and possibly changes in thought patterns (e.g., personality),[10] yet many high-level athletes want anything that might give a competitive edge.

Many other situations, besides sports competitions, lend themselves to a desire to work through pain or tiredness. Some claim that tDCS can also improve overall thinking. Search up tDCS, and many advertisements for brain stimulating devices–even travel sized!–show up. One web site boasts:

> tDCS can increase cognitive performance on a variety of tasks, depending on the area of the brain being stimulated. Scientific studies have shown that tDCS has the ability to enhance language and mathematical ability, attention span, problem solving, memory, and coordination. In addition, tDCS has also been documented as having impressive potential to treat depression, anxiety, PTSD, as well as chronic pain.[11]

While these more enthusiastic claims about tDCS are debated, there is increasing scientific evidence that functional magnetic resonance imaging (fMRI) neurofeedback may increase attention and visuospatial memory.[12] Another low-risk brain stimulation technique is electroencephalograms (EEGs). EEGs are used, for example, to stimulate an athlete's motor learning and monitoring motor function through the reading of biomarkers that predict athletic performance.[13] These biomarkers provide feedback on sleep, stress levels, focus, and impulse control.

[8] https://www.ftc.gov/enforcement/cases-proceedings/refunds/geniux-refunds.

[9] A. Hutchinson, *Mind, Body and the Curiously Elastic Limits of Human Performance* (New Zealand: HarperCollins Publishers, 2018).

[10] L. Y. Cabrera, E. Y. Evans, and R. H. Hamilton, "Ethics of the Electrified Mind: Defining Issues and Perspectives on the Principled Use of Brain Stimulation in Medical Research and Clinical Care," *Brain Topography* 27 (2014): 33–45.

[11] "The Brain Stimulator: Stimulate Your Life." https://thebrainstimulator.net/what-is-tdcs/.

[12] Dressler et al., "Hacking the Brain."

[13] G. Cheron et al., "Brain Oscillations in Sport: Toward EEG Biomarkers of Performance," *Frontiers in Psychology* 7, no. 246 (2016). https://doi.org/10.3389/fpsyg.2016.00246.

Our Brains on AI

If we choose not to tamper with our own physiologies, AI is the most obvious and fast developing way to augment thinking. AI offers many health benefits, such as diagnosing sleep disorders from home and monitoring cardiac performance with wearable technology.[14] Other examples include electronic memory aids and hand-held memory-enhancing digital games. The list is expanding.[15]

Developments are occurring regularly in the field of neurotechnology. Notably, brain-computer interfaces are increasingly enabling people to control devices with their brains. Building on the work of pioneers like Kevin Warwick,[16] CTRL-Labs is a United States wearable tech company building technology that allows for control of digital devices with the brain. Facebook acquired CTRL-Labs and joined it with Facebook Reality Labs, giving the effort enormous funding. Another neurotechnology company, Neuralink, is backed by billionaire Elon Musk, an indication that neurotechnology is supported by powerful individuals and organizations.

The Internet of Bodies (IoB) is an extension of the Internet of Things (IoT), which is comprised of interrelated mechanical and digital machines transferring data without regular human assistance. The IoB is connected to the IoT via devices implanted, ingested, or worn. Basically, the human body is used as a data platform. The IoB can augment individual cognitive abilities by giving us more information about ourselves and others and by interpreting that data.

For example, smart pills with electronic sensors and computer chips collect data about internal organs as they make their way through the digestive tract. Pacemakers now have Wi-Fi capacity and send data about heart function to a computer. Biohax, a Swedish bioengineering company, implants microchips (biochips) into bodies to enable people to enter their workplace without an external key and to pay for purchases simply by waving their hand.[17] The increased interfacing of humans with machines supplements our cognitive capacities by adding machine collected IoB data and by applying a fast-growing body of algorithms through which to interpret this data.

[14] Isabel Pederson and Andrew Iliadis, eds., *Embodied Computing—Wearables, Implantables, Embeddables, Ingestibles* (Boston: MIT Press, 2020).

[15] For a visionary discussion of what could be coming in brain enhancement, see this examination by theoretical physicist Michio Kaku, *The Future of the Mind: The Scientific Quest to Understand, Enhance, and Empower the Mind* (New York: Doubleday, 2014).

[16] Kevin Warwick, "The Cyborg Revolution," *NanoEthics* 8 (2014) 263–273. 10.1007/s11569-014-0212-z.

[17] "Biohax International: Turning the Internet of Things into the Internet of Us." https://www.f6s.com/biohaxinternational.

Technology of Moral Enhancement[18]

As we greatly enhance our cognitive reach, our capacities for both good and bad are amplified. Given the potentially nefarious outcomes perpetrated by superintelligent machines or radically cognitively-enhanced and supplemented humans, a reasonable fear is that unless we become more moral and collectively committed to doing good, we may act on our increasing capacity to do great harm.

Moral enhancement may be a necessary complement to the increasing potency of cognitive enhancements. Ingmar Persson and Julian Savulescu, Director of the Oxford Uehiro Centre for Practical Ethics, published *Unfit for the Future: The Need for Moral Enhancement*,[19] a book regarded as the pivotal text on moral enhancement ethics. Most scholars responding to their arguments see the moral enhancement project as too complex, fraught with danger, and unlikely to be implemented. But the fact that their book has been central to most analyses of moral enhancement suggests that Persson and Savulescu struck a deep chord.

Persson and Savulescu argue that we may be able to increase our capacities for altruism and justice and that we must do so or risk a cataclysmic fate. Without a stronger moral compass and inclination, we will be at greatly increased risk of mass destruction. Consider, for example, autonomous weapons, such as self-flying drones with sniper sensor devices and facial recognition software that tell these drones what skulls to penetrate and how to evade bullets. These unstoppable "slaughterbots" are not merely science fiction. This can be done now by integrating technology we already have in miniature form.[20]

Our moral historical track record is not good. Millions of people have been bombed, genocides have unfolded without adequate intervention, and people continue to be persecuted simply for being different. The world displayed plenty of altruism during the COCID-19 pandemic, but the pandemic also showed how territorial, greedy, and self-centered people can be. Perhaps the answer does lie in making humans more moral or more virtuous. But, how do we do this? This "how" question is what elicits strong reactions and complicates the ethical project.

[18] Much of the material in this chapter on moral bioenhancement is an adaptation from Tracy J. Trothen's publications, including "Moral Bioenhancement Through An Intersectional Theo-Ethical Lens: Refocusing on Divine Image-Bearing and Interdependence," *Religions* 8, no. 5 (2017): 1–14, 10.3390/rel8050084; and "Moral Bioenhancement From the Margins: An Intersectional Christian Theological Reconsideration," in *Religion and Human Enhancement: Death, Values, and Morality*, eds. Tracy J. Trothen and Calvin Mercer (Hampshire, UK: Palgrave Macmillan), 245–263.

[19] Ingmar Persson and Julian Savulescu, *Unfit for the Future: The Need for Moral Enhancement* (Oxford, UK: Oxford University Press, 2012).

[20] "Slaughterbots," last modified November 12, 2017. https://www.youtube.com/watch?v=9CO6M2HsoIA.

How We Make Ourselves More Moral

Traditionally, we have relied on communities and families to instill a moral compass in each person. Education and formation of a moral code through hands-on experience, reading, and debriefing after moral struggles have been some of the primary ways we have engendered and cultivated morality. Historically, religious communities have played a key role in instilling a moral compass and promoting the virtues. The religious traditions have long held that discipline, study, and participation in a faith community are vital to the development of morality.

Critics rightly wonder about the capacity of neurobiological agents to make us more virtuous, even potentially taking the place of communities, families, and religion. The development of a moral, virtuous person has long been understood as requiring time and discipline. Some view a quick fix dispensed at a biological level as repugnant, seemingly reducing humans to mere machines to be repaired. Indeed, moral enhancements are usually referred to as moral bioenhancements, because most interventions are biomedical. If morality cannot be reduced to biomedicine, then these bioenhancements have limited value. However, some see this limited value as potentially very important, and others do believe that morality can be reduced to biomedicine.

In the coming decades, conscience apps and morality software may well constitute a bridge between traditional methods of moral development and the more radical interventions being contemplated.[21]

Morality in a Pill?

Morality is influenced by neurobiology and so, potentially, can be affected by drugs and other interventions that change our nervous system. For example, there is behavioral, genetic, and neuroscientific evidence that aggression has a biological basis.[22] Such a finding begins to lay the foundation for developing moral bioenhancement programs. Numerous pharmaceuticals are already candidates for such programs.

The drug Ritalin reduces impulsive aggression. Ritalin can also sharpen one's ability to focus and problem solve more deliberately, even about ethical questions. The drug Provigil (modafinal) may increase prosocial behaviors, such as empathy, cooperation, trust, and concentration. The hormone serotonin increases aversion to harming others and increases empathy. The hormone oxytocin increases prosocial behaviors, such as empathy, cooperation, and trust.

In some situations, we may consider more aggression to be morally better. In highly competitive sports, for example, confident aggression is often seen as

[21] James Hughes, "How Conscience Apps and Caring Computers will Illuminate and Strengthen Human Morality," in *Intelligence Unbound: The Future of Uploaded and Machine Minds*, 26–34, eds. Russell Blackford and Damien Broderick (West Sussex, UK: Wiley Blackwell, 2014).

[22] T. Douglas, "Moral Enhancement," *Journal of Applied Philosophy* (2008) 25: 233.

morally virtuous in athletes, so long as it does not lead to undue violence. So, the heightening of aggressive impulses may be morally enhancing, and we can indeed heighten aggressive impulses with central nervous system stimulants, such as methylphenidates, ephedrine, and amphetamines.

These pharmaceuticals carry risks. Oxytocin can make people more trusting, but it is not advisable to be more trusting in all situations. Oxytocin increases altruistic behavior and empathy but only towards people we see as close to us or as kin. So, oxytocin may bring us closer to kin but might make us more distant from and suspicious of others. Maybe increasing some prosocial behaviors and decreasing aggression does not in total enhance morality in all situations.

Brain Stimulation

Currently, pharmaceuticals are the most promising avenues for moral enhancement. However, brain stimulation is also a pathway. Brain stimulation was developed mostly for the treatment of some diseases, including Parkinson's Disease and major depression. Transcranial direct current stimulation was designed for the treatment of major depression, but tDCS could also be used as a moral bioenhancement since it may increase cooperation[23] and neuroplasticity, making it easier in general to learn and, in particular, easier to learn prosocial behaviors.

Whatever moral benefit comes with brain stimulation, as we pointed out earlier, that benefit is not risk free. Deep Brain Stimulation (DBS), Transcranial Stimulation (TMS), and tDCS all can cause seizures or headaches. Perhaps more concerning, they may affect personal identity in unforeseen ways by changing thought patterns. Personality change certainly constitutes a major risk.

Genetic Modification

Genetic modification technologies will likely be marshalled to increase or decrease certain behaviors or thinking patterns in the direction of greater morality. For example, a protein called the Downstream Regulatory Element Antagonistic Modulator (DREAM) is associated with how we experience pain sensations. The protein could theoretically be edited out to increase pain toleration by blocking or dampening pain sensations.[24]

On first glance, pain may not seem to have much to do with morality. But pain sensations can tell us to stop a behavior causing the pain or to get help. Sometimes our experience of physical pain helps us learn what is harmful to

[23] A. Piore, "A Shocking Way to Fix the Brain," *MIT Technology Review* (2015). Accessed February 3, 2017. https://www.technologyreview.com/s/542176/a-shocking-way-to-fix-the-brain/.

[24] A. Miah, "The DREAM Gene for the Posthuman Athlete: Reducing Exercise-Induced Pain Sensations Using Gene Transfer," in *The Anthropology of Sport and Human Movement: A Biocultural Perspective*, eds. R. R. Sands and L. R. Sands (Plymouth, UK: Lexington Books, 2010), 327–341.

ourselves and to others. Without pain sensations, we are at greater risk of harming ourselves and of being insensitive to and unaware of pain others experience. From this perspective, editing out the DREAM gene would be morally detrimental. Yet, from another perspective, the reduction of pain could embolden us to pursue worthy, albeit physically demanding goals, such as working harder to win an Olympic medal. In a military context, soldiers could better overcome injury or exhaustion and more effectively defend and protect others.

Another example of possible future moral enhancement using genetic modification is the reduction or even elimination of fear and traumatic memory formation.[25] The military, for example, would certainly have interest in such neuroscience research. Such a moral bioenhancement could also function as an affective enhancement, to be discussed in the next chapter.

Robotics

We are extending, or supplementing, our moral and affective reach through AI. AI robots now provide comfort in hospitals and can even perform some duties provided by clinical professionals. During the COVID-19 pandemic, "robot pet therapy" was used to comfort elderly hospital patients who had very limited physical contact with their family members and friends.[26] Pepper is a robot with a humanoid appearance created by Softbank Robotics in Tokyo, Japan. The robot interacts with patients and their families at Humber River Hospital in Toronto. Equipped with sensors and cameras, Pepper has the ability to detect emotions and respond to people in prosocial ways. Pepper's prosocial example may help teach moral behavior, in addition to improving our emotional well-being. AI is providing us with new ways to express the virtue of caring.

Robots seem to have much to offer, but there may be limitations having to do with relationship and human touch.[27] As the COVID-19 pandemic has shown us, people in pain and frightened for their lives and their loved ones, want human touch and human presence. Spiritual distress heightens our need for physical human contact and accompaniment. A robot such as Pepper may be a helpful moral and emotional adjunct to a person, without replacing the human agent.

[25] M. N. Tennison and J. D. Moreno, "Neuroscience, Ethics, and National Security: The State of the Art," *PLOS* (2012). http://journals.plos.org/plosbiology/article?id=10.1371/journal.pbio.1001289.; and R. K. Pitman et al., "Pilot Study of Secondary Prevention of Post-Traumatic Stress Disorder with Propranolol," *Biological Psychiatry* 51, no.2 (2002): 189–192.

[26] Kate Knibbs, "There's No Cure for Covid-19 Loneliness, but Robots Can Help," *Wired Magazine* (June 22, 2020).

[27] Corinne Purtill, "The Robot Will Help You Now: How They Could Fill the Staffing Gaps in the Eldercare Industry," *TIME Magazine* (November 4, 2019).

Empathy Labs

Ongoing research suggests we can learn empathy to at least some degree, developing increased sensitivity to experiences and emotions of others. Altruism, which is closely related to empathy, is one of the two main virtues promoted by Persson and Savulescu in their argument for moral enhancement. Simply put, altruism is about selfless actions directed at the well-being of others, and a significant dimension of empathy is the ability to understand how someone might be feeling in a given situation. Empathy can help, and may sometimes be necessary for, people to behave altruistically. Experiential programs, such as role-playing and simulation exercises, are emerging as the most effective ways to teach the cognitive domain of empathy. Empathy "labs" have used such teaching strategies with encouraging results.[28]

It is thought that empathy has three domains: cognitive, affective, and behavioral.[29] Of the three, the most success has been in teaching cognitive empathy,[30] "the ability to know and understand that other people have a diversity of perspectives that are informed by thoughts and emotions that may be similar to or different from our own."[31] These new and emerging teaching techniques are non-biological moral enhancers.

As is likely becoming apparent, enhancement categories sometimes overlap, because different aspects of being human cannot be neatly separated. One of these overlaps is between moral and spiritual enhancement. Spirituality is associated with increased empathy, compassion, and altruistic behavior toward strangers. Unlike oxytocin, increased spirituality does not heighten altruistic behavior and empathy *only* towards people who we see as close to us, such as friends and kin, but also towards strangers.[32] So, spiritual enhancement means may also be morally enhancing. We address spiritual enhancement in the next chapter.

[28] Linus Vanlaere, Trees Coucke, and Chris Gastmans, "Experiential Learning of Empathy in a Care-Ethics Lab," *Nursing Ethics* 17, no. 3 (2010): 325–336; Linus Vanlaere, Madeleine Timmerman, Marleen Stevens, and Chris Gastmans, "An Explorative Study of Experiences of Healthcare Providers Posing as Simulated Care Receivers in a 'Care-Ethical' Lab," *Nursing Ethics* 19, no. 1 (2012): 68–79.

[29] G. Ançel, "Developing Empathy in Nurses: An Inservice Training Program," *Archives of Psychiatric Nursing* 20, no. 6 (2006): 249; Vanlaere, et al., "An Explorative Study," 70; and S.A. Batt-Rawden, M.S. Chisholm, B. Anton, and T. E. Flickinger, "Teaching Empathy to Medical Students: An Updates, Systematic Review," *Academic Medicine* 88, no. 8 (2013): 1171.

[30] Batt-Rawden et al., "Teaching Empathy," 117.

[31] Tracy J. Trothen, "Moral Bioenhancement Through An Intersectional Theo-Ethical Lens: Refocusing on Divine Image-Bearing and Interdependence," *Religions* 8, no. 5 (2017): 1–14. 10.3390/rel8050084.

[32] Laura Rose Saslow, et al., "The Social Significance of Spirituality: New Perspectives on the Compassion–Altruism Relationship," *Psychology of Religion and Spirituality* 5 (2013): 201–18.

Religious Issues

Religions Agree on the Goals

Wisdom

Intellectual development is important in all religions. Although some more than others, every religion has a long philosophical and intellectual history. Teaching and learning are central missions of temple, church, mosque, and ashram. Most religious leaders are charged with a teaching mission.

Islam, for example, welcomes science. The Prophet Mohammad called scholars the heirs of the prophets,[33] and it is obligatory for every Muslim to acquire knowledge.[34] The proviso is that scientific knowledge, as with any knowledge, must help to bring one closer to God through the pursuit of good works that reflect the valuing of each person. While there are plenty of instances, in past and present times, of the monotheistic religions resisting science in favor of religious ideology, all religions to a significant degree, monotheistic and karmic, have played a role in support of intellectual and scientific enterprises.

It is important to distinguish general knowledge from wisdom, a very different and special kind of knowledge. "Wisdom" books actually constitute a genre of literature in the ancient New East, the cultural context giving birth to the monotheistic religions. The "high" or philosophical wisdom books teach deep truths about perennially difficult topics, such as suffering, virtue, and the meaning of life. One of the highest Israelite virtues, wisdom, is personified as a righteous woman in ancient Jewish scriptures. In Buddhism, wisdom that allows one to see the true nature of things is liberating.

So, wisdom in the religions, which are sometimes called the "wisdom traditions," is certainly not reduced to intellectual attainment. Wisdom goes far beyond cognitive abilities like memory and processing speed and far beyond the accumulation and processing of data. Wisdom entails insight, judgment, and self-knowledge. Drawing upon the chapter, "Radical Human Enhancement and Ethics," wisdom, revered in the religions, entails self-awareness and self-reflexivity. Wisdom informs good ethics.

Morality

The religions also agree on the importance of living a moral life. They all articulate the particulars in varying ways, and frame their moral codes differently, but there is an interesting similarity among the religions in this regard. They all assign importance to being moral in the world. The theme of compassion, animating a moral life, runs through the sacred texts and teachings of the religions.

The ten commandments are central to Judaism and Christianity. In the Christian tradition, it is believed that Jesus knew the ten commandments,

[33] Al-Kulayni, vol.1, 39.

[34] Mohammad-Baqer Majlesi, *Bihar al-Anwar*, vol.1, 177.

embraced them, and gave them his own interpretation. The commandments value life, property, truth, and commitment. Shi'a Islam, a major branch of Islam, teaches the principle of *Adl* (Arabic, "justice"), which includes the conviction that God acts based on a divine design or plan and that God gives people the necessary instruction to know the difference between good and bad and to choose good. Jurisprudent schools of thought in the Sunni Muslim branch differ, but they all agree the believer is obligated to a moral code that guides behavior.

Hinduism instructs the faithful to follow the *yamas*, the five abstentions. Do not harm, lie, steal, indulge, or covet. The five Hindu *niyamas*, the five observances, are purity, contentment, discipline, study, and commitment to God. Buddhism's eightfold path includes right speech, effort, and conduct. Buddhists love lists, and right conduct includes the five precepts of refraining from killing, stealing, lying, unchastity, and intoxication.

Driving the moral energy of the religions are love and compassion. The monotheistic religions and the karmic ones generally unite in giving attention to compassionate service to others.

How to Get There

Traditionally, wisdom and morality are achieved through study of sacred texts and doctrinal training, starting ideally at an early age and continuing through adulthood. Religions tend to be family-oriented traditions, and religious training also, ideally, occurs partly in the home. The monastic traditions, which some religions have, are exceptions to the family orientation.

While every active religious follower is trained in wisdom and moral formation, religious leaders get special and extended training. The religions and the denominations within the religions have different educational standards and protocols. However, they generally value leaders having knowledge of sacred texts, doctrine, ritual, and practice. In Hinduism and Buddhism, the guru-disciple, i.e., the teacher-student, relationship is central to maintaining the integrity of the tradition. This teacher-student relationship is absolutely essential in traditional Zen Buddhism.

The point we are making is that much time and effort goes into transmitting wisdom and forming moral character in the religions. In this chapter we have seen that radical cognitive and moral enhancement potentially can be at least partially achieved without the time and effort needed in the past. Technology and pharmaceuticals are likely to offer society the option of cognitive enhancement, in addition to traditional education delivery systems. Smart pills and moral bioenhancements likely will be available to multiple segments of society, including the religions.

There is, of course, a question about how effective cognitive and moral enhancements will be. If they prove effective, a follow-up question for the religions is whether the wisdom, knowledge, and morality instilled by the pill or brain stimulation is the same as that achieved through long, disciplined study and expert guidance from religious leaders, or if these enhancements might be helpful supplements to religious programs and formation. Our goal here is to describe the issues likely to be faced by the religions, not to answer the questions.

Another issue for the religions to consider is the importance of effort. In the karmic religions, good effort helps determine one's status in the next rebirth. If effort is effectively eliminated with a pill, that poses an interesting question about the nature and role of karma. Right effort is one of the tenets of the Buddhist eightfold path. Perhaps Buddhism, and the other religions, will determine that discipline and effort are essential features of the journey to wisdom and a moral life. If so, the religions may reject the quick fix, shortcut enhancement. Or, maybe the religions will see these radical enhancement possibilities as opportunities to achieve wisdom and morality quicker, enabling followers to more faithfully and effectively live the devout life.

It is important to understand that we are only at the beginning of cognitive and moral enhancement technological possibilities. The pills and brain stimulation methods today will likely look primitive decades from now. For example, perhaps at some point we will be able to generate a neuro-biomedical profile of each person, informing how best to adjust certain neurotransmitter and other biochemical levels more optimally for each person, and to select what virtues to enhance based on each individual's particular needs. Buddhist transhumanist James J. Hughes suggests this.[35] So, the religions, and society, may not need to decide for or against a particular moral bioenhancement in general but, rather, for or against a particular moral bioenhancement for each person.

Finally, if the religions utilize technology to advance their teaching and moral missions, the religions will need to critically examine that technology. Importantly, AI programmers employed by leading technology companies are overwhelmingly not people of color and do not have first-hand experience of what it is like to live with a skin color other than white-appearing.[36] If AI is partly informed by racialized attitudes, then we can expect that AI will amplify already existing racialization. We will revisit this issue in the ethics section of this chapter.

Self-reflexivity shows us that our context and embodied experiences have a big impact on how we see things or do not see things. So, while AI may help us to make better use of our existing knowledge and to develop more knowledge, AI may also increase the disadvantages experienced by marginalized

[35] James J. Hughes, "Ancient Aspirations Meet the Enlightenment," in *Religion and Human Enhancement: Death, Values, and Morality*, eds. Tracy J. Trothen and Calvin Mercer (Hampshire: Palgrave Macmillan, 2017).

[36] Stephanie Dinkins and Charlton Mcllwain, "Coding While Black: Artificial Intelligence, Computing, and Data in a Racialized World," *Initiatives Emerging Leaders Program Blog Humanizing Data Review* (New York University, 12 March 2018). https://urbandemos.nyu.edu/2018/03/12/coding-while-black-artificial-intelligence-computing-and-data-in-a-racialized-world/. https://urbandemos.nyu.edu/2018/03/12/coding-while-black-artificial-intelligence-computing-and-data-in-a-racialized-world/.

groups. Or, it could be that some cognitive enhancements will improve creative thinking and will help us to better consider a wide variety of perspectives. The principle of self-reflexivity means that context and embodied experiences impact our perspectives. The religions will want to ensure the therapies and technologies they employ for religious goals reflect values congruent with religion.

The Theological Continuum

As with most enhancements, there will likely be a number of approaches and positions taken by religious adherents, with regard to the potential usage of cognitive and moral enhancements.[37] Given the many branches and denominations in each religion, and many followers, we will surely find varying assessments of cognitive and moral enhancement. Referring back to the theological continuum in the chapter, "Transhumanism, the Posthuman, and the Religions," more conservatives may tend to reject enhancement technologies, while more liberals may lean toward acceptance.

Conservatives are traditionalists, tending to do things as they have always been done. Given the central role of family and religious institutions in the training of children and adult converts, conservatives may find it difficult to hand these tasks over to what they may interpret as secular, science-based methods. Conservatives may see the use of external substances, such as pharmaceuticals, as unacceptably tampering with God's good creation. It may be that over time, some of these enhancements become accepted as normal among conservatives, but studying holy scripture and the faith community's doctrines probably will remain a central way of becoming smarter, wiser, and more virtuous. Following these habits and disciplines, it is believed, will help conservative followers get into heaven after death in the monotheistic religions and reborn into a better status in the karmic religions.

Liberals will certainly have reason to be wary, but on the matter of tradition, they tend to be revisionists, open to science and willing to utilize modern methods, replacing old ways that no longer work well. However, liberals will have other considerations, such as fair access, and distributive and social justice, that could prompt them to resist these enhancements, as well as enhancements explored in other chapters.

We explained how wisdom is distinguished from the mere learning of facts or information. Cognitive enhancement could be useful in the pedagogical mission of the religious institutions. Moral enhancement could be useful to the religions as well. But we can anticipate that the religions may choose to reserve the teaching of the deep wisdom of the religion for parents and trusted leaders. We are speculating on all these possibilities. Radical cognitive and moral enhancements are still largely in the future, so we will see in the coming years how the religions assess and utilize, or not, these technologies.

[37] A range of positions can be found, for example, in this theme issue of *Theology and Science* 16/3 (2018), devoted to "Moral Enhancement and Deification through Technology?"

Tower of Babel

An oft-told biblical narrative in the Jewish and Christian traditions is the story of the Tower of Babel in the Hebrew Bible's book of Genesis.[38] Traditional interpretation holds that the human beings acted irresponsibly in the Garden of Eden,[39] Cain killed Abel,[40] and the wickedness was such that the human community brought upon itself a catastrophic flood.[41] In the tower story, perhaps out of fear of being "scattered abroad" or because of the narcissistic impulse to "make a name for ourselves," the human community proposes to use their available technology to build a tower "with its top in the heavens."

This old story can lend itself to various interpretations, one of which is that the story depicts the creature's attempt to "be like God,"[42] to use a phrase from an earlier story in Genesis. Or, to put it more bluntly, to *be* God. In this interpretation, the story seems particularly directed at those with the most social and economic power and who have the capacity to assert and implement that power in widespread political ways. Striving to reach one's potential—to fulfill one's God-given purpose—is one thing and is quite appropriate. However, in the monotheistic theological model, from which the Tower of Babel story comes, human beings are a part of the created order. They are not God. They are not omniscient and should not strive to be so. There are appropriate limits to who and what human beings are meant to be.

Seeing the tower being built, God said, "This is only the beginning of what they will do; and nothing that they propose to do will now be impossible."[43] In light of some of the dazzling enhancement technologies under way, such as the cognitive and moral ones addressed in this chapter, this old text about everything being possible takes on a new relevance. The attempt of the more systemically powerful humans to build the tower to heaven did not end well. The people's language was confused, and they were scattered abroad over the face of all the earth. Perhaps a warning is found here for those who desire to know and understand everything via cognitive enhancement and achieve divine moral perfection with pharmaceutics.

In a similar vein, the religious notion of sin may not be a quaint, outdated idea for the world of radical enhancement technology. The notion that human beings have tendencies toward self-serving, even hateful and hurtful, behavior can be a caution to society about rushing uncritically into every new technology. It is certainly appropriate for the religions to take human capacity for depravity, however interpreted, seriously as these therapies and technologies grow.

[38] Genesis 11:1–9

[39] Genesis 3.

[40] Genesis 4.

[41] Genesis 6–9.

[42] Genesis 3:5.

[43] Genesis 11:6.

While caution is in order, the religions also have long traditions and religiously based moral imperatives to do good. Improved cognition could theoretically help religious followers do even more good. The challenge religious followers will face is being faithful to their religious commitments in the context of deciding what role enhancement technologies can and should play.

Perhaps charting a path many religious followers will take, Persson and Savulescu contend that if moral bioenhancements such as oxytocin, which enhance only in-group empathy, are to be effective, traditional methods of education must continue: "moral bioenhancement would have to go hand in hand with reasoning which undercuts race, sex, etc. as grounds for … differentiation."[44] Moral bioenhancements are very unlikely to cut out the need for traditional moral education and formation, but moral bioenhancements may eventually be able to step up human capacity to be more moral.

Will Religious Leaders Be Out of Work?

At risk of being cynical, the role and job security of religious leaders could, conceivably, play a role in how they lead the religions to respond to cognitive and moral enhancement. Religion has psychological and sociological dimensions. And, religion is also about economics. The pastors, priests, imams, and swamis of the religions play central roles in teaching wisdom and guiding the adherents in forming a moral life.

Certainly, most leaders are called to service and committed to their work as a sacred mission. They are also invested in their work as a profession and are usually paid salaries in some form or other. If technology should develop that can accomplish the religious leader's teaching mission, they are then out of that job. Understandably, there could be resistance to this development. That resistance may be framed theologically or ethically, but the reality is that, consciously or unconsciously, economics could play a significant role.

However, even if cognitive and moral enhancements develop to the point at which we are not as dependent on traditional methods of moral and cognitive growth, presumably religious leaders would still be needed to lead other aspects of faith communities. And, as we suggest above in tandem with Persson and Savulescu, technology enhancements may provide a huge boost but will not likely provide us with all that we need to become more moral and smarter, especially if by smarter we also mean wiser.

Ethical Issues

Ethical and religious issues cannot be neatly divided. Faith claims inform how followers act in the world and how they will think about radical enhancements. So, we will see ethics threaded throughout our reflections on religion, and we

[44] Ingmar Persson and Julian Savulescu, "The Evolution of Moral Progress and Biomedical Moral Enhancement," *Bioethics* 33, no.7 (2019): 816.

will see religious issues threaded throughout our reflections on the ethics of cognitive and moral enhancement. Remember, too, we are only scratching the surface of these issues. As always, consider some of the sources provided in the footnotes if you are intrigued and want to learn more about an issue.

Cognition and Morality: What Exactly Are We Trying to Enhance?

Beyond the potential bad side effects of the interventions, two important questions loom with cognitive and moral enhancements: What is meant by cognition, and what is meant by morality? As discussed earlier in this chapter, cognition involves several domains of ability, and a single enhancement is not—at least so far—going to enhance all cognitive domains.

It is more complicated than that. It is not even clear that cognition is limited to the brain. We have body memories. For example, the heart beats faster before the brain realizes something in the environment is stirring up bad memories. Perhaps body memory is a cognitive domain. If so, if we enhance the brain's capabilities, and leave the body behind, perhaps we are not enhancing one's cognition with integrity. Artists and athletes commonly experience their bodies as leading their brains in artistic and sport activities.

Morality is at least as complicated. Morality is understood in a variety of ways, including (1) the discipline of being and becoming virtuous, (2) the capacity to make sound and well-reasoned ethical arguments, and (3) acting in ways that conform to the virtues. Some theorists focus on how we might enhance moral behaviors, while others are more concerned with enhancing the motives behind behaviors.[45] Still others think we can or should enhance universal virtues, such as restraint, altruism, and inclination toward justice.[46] Others are more hopeful that we might enhance cognitive abilities, so that we improve our capacity for ethical reasoning.[47] Your authors believe the main reason morality is difficult to define is its contextuality. The meaning of doing good and of virtues, such as altruism and justice, changes a little or a lot depending on who is interpreting what it means to do good, and the virtues.[48]

Not all prosocial qualities are desirable in all moral situations; there are times when the moral road is to be aggressive or even violent.[49] For example, you see a car coming quickly towards someone. Violently pushing them out of the way

[45] Douglas, "Moral Enhancement."

[46] Persson and Savulescu, *Unfit for the Future*; and James J. Hughes, "How Moral is (Moral) Enhancement? Moral Enhancement Requires Multiple Virtues Toward a Posthuman Model of Character Development," *Cambridge Quarterly of Healthcare Ethics* 24 (2015): 86–95.

[47] John Harris, "Moral Enhancement and Freedom," *Bioethics* 25 (2011): 102–11.

[48] Michael Hauskeller, "The Art of Misunderstanding Critics: The Case of Ingmar Persson and Julian Savulescu's Defense of Moral Bioenhancement," *Cambridge Quarterly of Healthcare Ethics* 25 (2016): 153–61.

[49] D. Gareth Jones, "Moral Enhancement as a Technological Imperative," *Perspectives on Science and Christian Faith* 65 (2013): 150; Inmaculada De Melo-Martin and Arleen Salles, "Moral Bioenhancement: Much Ado about Nothing?" *Bioethics* 29 (2015): 223–32; and Hauskeller, "The Art of Misunderstanding Critics."

may save them from an oncoming vehicle. Consider an all too common bullying situation. When a bully harasses someone, the bully needs to be challenged, not accommodated. Each situation is different, which is a key point addressed by situation ethics and consequentialist ethics, introduced in the chapter, "Radical Human Enhancement and Ethics."

The virtues we need depend a lot on the situation and on our personal and social identities, as we shall see. We do not all necessarily need the "upping" of all virtues, such as generosity, altruism, empathy, justice, courage, faith, hope, and charity. Some may need prudence first and perhaps only. Prudence is wisdom and good judgement. And, we need the will or motivation to do good. Suffice it to say, for now, that what we mean by cognition and morality influences our understanding of what it means to become cognitively or morally better.

Values, Virtues, and Self-Reflexivity

As explained in the chapter, "Radical Human Enhancement and Ethics," values are those things that are most important to us. Virtues are traits or qualities considered morally desirable. So, a close relationship usually exists between what we see as virtues and the things we value. For example, if faith is perceived as a virtue, then we might value church. Or, if we see altruism as a virtue, we likely value friends and strangers regardless of who they are.

To illustrate, the spouse of one of your authors, Professor Trothen, recently proceeded through the drive-through at a Tim Horton's, a donut and coffee/tea chain close to the hearts of many Canadians. When he arrived at the window to pay for his steeped tea, he was told that the stranger in the car in front had already paid for him! If instances like this lift your spirits and move you, it may well be that you see altruism as a virtue, and value people and caring actions.

Very often we do not consciously know what we value most and what we see as virtues. Think again about your top five values, from the exercise in the chapter, "Radical Human Enhancement and Ethics." As discussed in that chapter, our values are at least partly determined by our contexts and our stories. What you value most may well be different from someone in another part of the world or from someone who has not had the same privileges as you, such as running water, electricity, and a stable home. Similarly, not everyone faces the same barriers, such as sexism, gender identity questions, ableism, racialization, and more. All these things affect what we see as most important, and everything in our lives has an impact on our values and on the virtues we learn and see as most desirable.

Because morality is contextual, we do not all need more of every virtue, even if we could target each virtue for enhancement. For example, we understand now that many women tend (but not always!) to be more self-sacrificing and nurturing. If we make someone who is already self-sacrificing, and maybe too much so, more self-sacrificing and altruistic, they may become a threat to their own well-being by not caring adequately for themselves.

Cognition is also contextual, and different types of cognition are valued differently depending on power patterns. In so-called mainstream North America, rational thought tends to be valued above—and regarded as separate from—relational or emotional ways of knowing. This valuing of rationality is derived from, at least in part, the power disbursal in this part of the world and the binary association of male gender with rational thought and female gender with relational or emotional knowing. So, it will not be a surprise when we talk more about values and cognitive enhancement in the section on justice later in this chapter.

The Therapy—Enhancement Continuum: What It Means to Make Us Better

As we discussed in the chapter, "Radical Human Enhancement Ethics," the therapy—enhancement continuum is a helpful, although imperfect, guide to considering which enhancements might be considered acceptable or unacceptable. One way religions could weigh in on an enhancement intervention is to judge it unacceptable at the point at which we cease to care for others and the earth and cease to prioritize good works.

Consider a follower of one of the religions who is cognitively enhanced so they can better focus, process thoughts, and think creatively. That, in itself, may be good, but if they are not more caring and concerned for others, or perhaps if they become even less concerned for others, then the enhancement may be judged undesirable or inadequate. Consistent with their values, the religions will likely want increased intellect to be accompanied by an increased capacity and willingness to do good, thereby becoming a better person. Thinking better is no guarantee we will behave better or be better.

Here is a psychotherapeutic clinical example of the above principle.[50] A young man with Asperger's syndrome who, not atypically for this disorder, tested well above normal on his Wechsler Adult Intelligence Scale IQ test. The youth was being seen for psychological assessment, because he had significantly compromised social skills. As it turned out, the youth's very high IQ compromised his ability to socially interact with his peers. Although very high IQs do not always pair with lesser social skills, it did in this particular case.

Increasing cognitive intelligence without a commensurate increase in social skills may very well result in an unhappy outcome, and diminished social skills is just one possible unintended negative consequence of significantly increased brain power. It may well be that cognitive enhancement (at least if it is without moral enhancement) leads to a widening gap between our intellectual abilities and our moral wisdom. Given the belief in the intrinsic worth and dignity of all

[50] This example, disguised so the client cannot be identified, comes from your author, Professor Mercer, who has worked as a therapist.

people, fostered by religions generally, thinking better will not necessarily make for better humans or trans/posthumans.

Perhaps good and safe moral bioenhancements can help to make us better people if we use them in conjunction with traditional methods for making us morally better. The word "good' is key. For all the reasons discussed so far, moral bioenhancements are not necessarily going to result in more moral people. As with cognitive enhancements, moral bioenhancements alone are unlikely to be sufficient to make us better. The religious traditions may very well insist that spiritual disciplines and education about self-reflexivity and justice issues must be combined with moral bioenhancements if the bioenhancements are to be effective.

Being human is about being more than a reducible collection of parts to be fixed with enhancement interventions.[51] People are more than enhance-able components and cranial neuro-pathways. Attempts to make us better morally may very well fall short if moral improvement is envisioned as a biochemical task only.

Choice

Who will choose to be cognitively or morally enhanced, and why will they make that choice? If it is a choice to be morally enhanced, we can imagine that most people opting for these enhancements may not need them as much as people who choose not to be morally enhanced. A conundrum! Perhaps we just legislate moral bioenhancements into tap water, much like we did with fluoride. That approach risks taking away people's ability to consent to becoming, potentially, a different person. It can be argued that authenticity requires that choice be protected, although some might take a utilitarian perspective and claim that the good of the many outweighs the value of preserving the choice to opt out. Deeply held values will play a role in deciding one's position.

The issue of choice is complicated by the question of whether our choices will still be our authentic choices after we are cognitively and/or morally enhanced. Maybe a morally enhanced person will not want any other enhancements, including those that let them live longer, unless everyone else in the world can have access to those enhancements. That may be admirable, but who is making the choice? Perhaps it is one's authentic self, but perhaps it is the bioenhancement making the choice, subverting authenticity. We use authenticity here to mean that one's choices and behavior reflect one's values and personal integrity. Will it be me making choices after I am cognitively or morally enhanced, or will I lose my authentic self by becoming changed?

Situating this conversation in theological language, will cognitive and moral interventions enhance or diminish the image of God, in the monotheistic

[51] For a thoughtful analysis of the mistaken tendency to see human beings as a series of reducible parts in relation to moral enhancement, see Harris Wiseman, *The Myth of the Moral Brain: The Limits of Moral Enhancement* (Boston: MIT Press, 2016).

religions, or change one's status in life, in the karmic traditions? A prominent view among ethicists is that a moral action must be freely chosen and not coerced in order for it to be moral. Ethicist John Harris has argued that we must have the freedom to fail if our actions are to have the potential of being authentically virtuous.[52] Will moral bioenhancements prevent us from freely choosing behaviors?

Consider the example of brain stimulation. TMS is described by Professor of neurology William P. Cheshire as

> a useful tool for investigating how specific brain correlates of self-awareness might be altered.... [which includes] an array of cognitive functions important to personal identity—moral reasoning, emotional valuation, decision making, unconscious bias, impulsivity, altruism, empathy, anxiety, fear, deception, belief and spirituality....TMS can influence assessments of threat or danger signaled by facial expressions. If visual perception and visual processing are subject to influence by technology then so might other brain capacities relevant to one's perception of others' or one's own personal identity.[53]

Brain stimulation may allow us to change ourselves, in the name of enhancement, to such a degree that we become almost unrecognizable.

A case can be made that we can be authentic only if we are not changed too much by technology or pharmaceuticals. Maybe these enhancements are inauthentic short-cuts. Or maybe enhancement technologies can help us become more of who we strive to be. Hughes, a Buddhist transhumanist whom we introduced earlier, argues that some of us have neurological barriers preventing us from being moral in ways that are authentic to ourselves. Perhaps a moral bioenhancement can remove a neurological barrier and allow someone to act in a more altruistic and less aggressive way that is more consistent with their authentic self.[54] If Hughes is right, then such bioenhancements are therapeutic, allowing some people to be more authentically human and true to their "normal" or authentic self. Of course, who decides and how they decide what it means to be authentic is, to say the least, difficult.

Here is yet another conundrum. Are our choices—even without being cognitively or morally enhanced—truly free? Consumers are bombarded by streams of media messages telling us what we supposedly really want. Everyone wants to sell us something, it seems. Maybe moral enhancements will actually let us become more of who we authentically are by overcoming neurological barriers to free choice.[55] Or, perhaps we will cease to make "bad" choices. For the religions, in a heavenly scenario, perhaps moral bioenhancement makes sin a thing

[52] Harris, "Moral Enhancement and Freedom,"103.

[53] William P. Cheshire, "Ethical Implications of Transcranial Magnetic Stimulation for Personal Identity," *Ethics and Medicine: An International Journal of Bioethics* 34, no. 3 (2018): 135–145.

[54] Hughes, "Ancient Aspirations."

[55] Hughes, "Ancient Aspirations."

of the past. Or not. Human beings are creating these bioenhancements, so it is very unlikely the bioenhancements will be perfect.

Justice

Choice is not just about the chooser. We cannot escape the reality that we are all connected. My choices affect you and vice versa. We cannot always know what these effects will be, but it is important to anticipate these effects when we approach cognitive enhancement and moral bioenhancement as justice issues.

Our values influence what we believe will make our thinking better and to make us morally better. People hold different values. Some values are relatively common, such as empathy and justice, although disagreements exist over their meaning and expression. Values, as we keep emphasizing, are socially influenced. Social processes affect what we think makes us better. For example, social processes influence the types of intelligence we most value. Certain types of intelligence, such as logic, are usually valued over other types, such as emotional intelligence.

Complicating things further, males and females are associated with different intelligences. Rightly or wrongly, males are more often linked with logic and females more with relational and emotional intelligence. The concern is that types of intelligence valued most in the current context will be enhanced at the expense of other intelligences, and that this valuing may be linked to the respective unjust valuing of different genders. Focusing, accumulating information, problem solving rationally, and memorizing may be emphasized at the expense of creativity, relational intelligence, musical abilities, symbolic thinking, intuition, and moral insight. How increasing some selected intelligences, and not others, will affect us is an important question.

Moral bioenhancements have limitations in addressing justice issues. Oxytocin increases empathy but only towards in-group members (e.g., kin). With the world being mired in ingroup/outgroup thinking, it is not clear that enhanced moral reasoning, even when combined with education regarding social justice, will be enough to make oxytocin more helpful than harmful. Ingroup/outgroup thinking has strong instinctual and emotional rootedness. Few people may be willing to do the self-awareness work needed to overcome such thinking.

It will be challenging and complicated to ensure that people who most need these enhancements get them, especially if they do not want them. From a utilitarian perspective, there is an argument to be made in favor of making proven moral bioenhancements compulsory. But forcing enhancements is, understandably, going to elicit resistance from many quarters.

If we fail to engage the variety of perspectives and especially perspectives of the socially vulnerable, we risk amplifying social inequities through enhancements. Consider the example of an elderly person with dementia who exhibits violent behavior, and there is a moral enhancement pill that could theoretically make them less aggressive. First of all, the decrease aggression pill may work on violent people without dementia but perhaps not on someone whose aggression is caused by dementia. The discussion could end here. However, perhaps it would be more appropriate and therapeutic to use moral bioenhancements to increase empathy and compassion in some clinical staff and managers of long-term care homes, rather than trying to fix the care-receiver. In other words, we need to explore different perspectives to assess who needs to become better and what might make them better. The principle of co-design means bringing in a diverse team with diverse perspectives. Co-design is increasingly important as "we" develop more ways to make "us" better.

Humanity needs moral improvement; there are few detractors on that point. If it turns out that morality can be enhanced safely, affordably, and justly, it is probably going to be most effectively done if it is done in addition to education. If, for example, it is agreed that prudence is important to acting and thinking morally, then prudence must be in the mix, either via a pill or from traditional methods, or both. Moral bioenhancements could be hugely valuable in today's world, but only if combined with knowledge about, for example, how systems disadvantage and privilege us. In other words, one possible happy future may involve the critical use of safe cognitive and moral enhancements combined with the deepest and best wisdom of the religious traditions.

Neither society, nor the religions, should give up on cognitive enhancement and moral bioenhancement just because they are really complicated and hard, which they are, both in terms of developing and ethically assessing. To give up would be failing to do as much good as reasonably possible. We need to keep working smartly, and one way is to find an appropriate balance between proactionary and precautionary approaches. We need to be cautious in the face of possible significant harms, and we also need to work to do as much good as possible while taking reasonable risks. The vulnerable should not have to bear the brunt of these risks.

We are often tempted by the easy fix. While radically increased intelligence, if properly managed, might be a good thing, we should not necessarily see it as a general solution for anything. We have solutions in hand for many of our societal ills; the problem is not in figuring out those solutions but in harnessing the will to implement the solutions. Too often, we would rather invest in a self-serving agenda.

Some people who are concerned about global warming point out that we already know how to address climate change. What is lacking is the political will and commitment to make major lifestyle changes. The COVID-19 pandemic showed us that we are indeed able to harness a mostly global will to work together to save as many people as possible, even at great economic cost. As with many challenges, it is not that we need scientists with higher IQ's, rather, we need large scale commitment to appropriate actions that we already know need to happen. Maybe moral bioenhancements can help. Maybe cognitive enhancements can help. But neither are likely to be sufficient by themselves.

Questions for Discussion

1. How might cognitive enhancement affect faith communities? Think of one example of a faith community and consider how cognitive enhancement might change worship practices. Will it change how followers act in their daily lives?
2. In Christianity and Judaism, it is believed that we are made in God's image. How far should followers go to be like God?
3. How might moral bioenhancements change how karma is understood?
4. Might moral bioenhancements be useful? From a theological perspective, do you think that these enhancements have limitations? Why or why not?
5. If we do not use moral bioenhancements, how do we safeguard against easy abuse and misuse of other enhancing technologies? What do you propose? Why?
6. Recall your top five values. How do they relate to your assessment of cognitive and moral enhancements? What do you think the religions might say about the relationship between values and making us smarter or more moral with enhancement technologies?
7. Discuss social justice as it relates to potential moral bioenhancements. If moral bioenhancements become available, should they be mandatory for everyone? Why or why not?
8. How would you assess the precautionary and proactionary calculation about cognitive enhancement? Moral enhancement?

CHAPTER 7

Affective Enhancement and Spiritual Enhancement: Feeling Happier and More Spiritual

Technology of Affective Enhancement

When basic survival needs of food, clothing, and shelter are met, human beings expend much time and energy on being happy and experiencing other positive emotional states. The world of commerce is filled with companies whose mission and products are designed to appeal to people's robust desire to feel good. Entire marketing campaigns are tailored accordingly. In the chapter, "Radical Human Enhancement and Ethics," we even mentioned an ethical theory, hedonism, that is oriented around maximizing pleasure and minimizing pain.

People are not always smart in the ways they strive for happiness, sometimes resorting to unregulated drugs, risky sexual encounters, and a host of other dangerous practices. The persistent reach for happy states suggests a strong market demand exists for more happiness, more good feelings. Given this demand, we can anticipate that therapies and technologies promising to enhance good feelings will get much attention in research, development, and commercial delivery. People can jump quickly into potentially risky behaviors in order to feel good and may very well do that with technology that promises to bring happiness. In this chapter we encourage the careful assessment of radical affective enhancements.

Making ourselves emotionally better is a project loaded with religious and ethical issues. We encounter questions similar to those asked with other enhancement categories, such as who gets to decide what makes one better in the emotional department, why some emotions are valued more than others, and who benefits from these enhancements.

How We Make Ourselves Emotionally Better

If the local pharmacy had a proven over-the-counter happy pill, without nasty side effects, we would probably find a long line at the counter, whether the pill

C. Mercer, T. J. Trothen, *Religion and the Technological Future*,
https://doi.org/10.1007/978-3-030-62359-3_7

is covered by insurance or not. In a thought-provoking piece, philosopher and ethicist Patrick D. Hopkins, an affiliate scholar with the transhumanist Institute for Ethics and Emerging Technologies, reflected on the meaning of salvation and the role of enhancements in the search for satisfaction. Based on historical evolutionary patterns, he does not think technology can deliver on promises to make us more satisfied in the long haul.[1]

Despite such reservations, the demand is there and, in this section, we consider some therapeutic and technological delivery methods, all of which hold the potential to be radically more potent in the future. And, on the horizon are breakthrough possibilities, such as genetic engineering. Before addressing the therapies and technologies that hold potential for radical biohacking, we provide a context by considering more traditional affective enhancement methods.

Traditional Affective Enhancement Methods

The attempt to enhance oneself affectively is not new. Long ago, Aristotle said people sought wealth, health, and friendships as means to achieve happiness. The desire to feel better emotionally has always been part of the human quest, and emotional improvement techniques continue to evolve. In the modern world, the profession of psychotherapy, with many modalities, arose to address dysfunction and help people heal emotionally. Cognitive Behavioral Therapy addresses distress or dysfunction through changing our thinking and our behaviors. Emotionally Focused Therapy is based on the premise that emotions are key. Narrative Therapy works with life stories, understanding people as separate from and bigger than their problems.

Given this chapter's topic of positive affect, "Positive Psychology" is an interesting approach, theorizing that boosting positive well-being pushes out negative emotion and gains the desired outcome of happiness or well-being. Martin Seligman's PERMA well-being is one of the better-known examples of positive psychology. Well-being is positively correlated with pleasure, engagement, relationships, meaning, and accomplishment, hence the acronym PERMA. Since about the turn of the century, there has been a good bit of interest in happiness research.

Spiritual care therapeutic approaches are concerned with accompanying people in their personal deserts, their bleak inner places, and exploring spiritual meaning or struggle in that context. At the heart of spiritual health approaches to emotional well-being is a desire to assist the person in achieving well-being, contentment, and possibly happiness. Modalities change and emerge in response to new research findings and the experiences that psychotherapists have with their diverse clients.

[1] Patrick Hopkins, "A Salvation Paradox for Transhumanism: Saving You Versus Saving You," in *Religion and Transhumanism: The Unknown Future of Human Enhancement*, eds. Calvin Mercer and Tracy J. Trothen (Westport, CT: Praeger, 2015), 71–82.

Pharmaceutical Agents

Mood altering drugs have been for some time an integral part of the life of people in modern, industrialized countries. One of your authors, Professor Mercer, worked as a therapist alongside psychiatrists who routinely drew upon a host of psychotropic medications, such as Prozac, Cymbalta, and Zoloft, to treat clients who wanted to be less anxious and less depressed. Mood can be managed by manipulating certain brain neurotransmitters, particularly serotonin, dopamine, and norepinephrine. Betablockers, such as Propranolol, can induce a non-anxious, calm, and focused state, which is why some archery and golf competitors use it. Testosterone is probably used by some athletes to feel more aggressive.

Neither the super happy pill nor "pick-the-emotion-of your-choice-pill" have yet to be developed. However, we are learning more about how the brain works, its role in emotional well-being, and how drugs can manipulate mood. Mood enhancement is a big industry now, and the demand is there for more radical measures. Current pharmaceutical mood interventions are likely to pale in comparison to what is coming.

Robots

Pills are not the only way enhanced affect is being packaged and sold. Consider these opening lines from a *New York Times* article, titled "Robots: Hot or Not? Love, Android Style, Sexy and Confusing":

> When Akihido Kondo, a 35-year-old school administrator in Tokyo, strolled down the aisle in a white tuxedo, his mother was not among the 40 well-wishers in attendance. "For mother," he told *The Japan Times*, "it was not something to celebrate." You can see why. The bride, a songstress with aquamarine twin tails named Hatsune Miku, is not only a world-famous recording artist who fills up arenas throughout Japan. She is also a hologram.[2]

"Digisexuals" refers to human-android romantic relationships. Online pornography, "hookup applications," sexting, and electronic sex toys constituted the first wave. A second digisexuality wave is opening up what practitioners testify to as deeper relationships through virtual reality, augmented reality, and artificially intelligent-equipped sex robots, programmed to be companions as well as sexual partners. Some observers are concerned we are headed for sex slaves, objectification, and fake relationships. Others welcome the new digital sexuality, expecting it to bring safety, enhance autonomy, and increase fulfillment. Whether technologically-enabled sexual and romantic relationships will enhance us affectively is a matter of debate and perspective.

[2] Alex Williams, "Robots: Hot or Not? Love, Android Style, Sexy and Confusing," *The New York Times*, *SundayStyles* (January 23, 2019): 1, 8. https://www.techvshuman.com/2019/01/23/do-you-take-this-robot/.

We will need to study the implications empirically, by conducting research including interviews, after the second digisexuality wave occurs.[3]

Cambridge University Professor and Computer Scientist Peter Robinson, an expert on affect and robots, reports that "it is becoming possible for machines to sense, analyze and express emotions."[4] We introduced Pepper, touted as "Canada's first emotionally sensitive robot for sick kids," in the chapter, "Cognitive Enhancement and Moral Bioenhancement." Pepper came to a Canadian hospital to help people feel better emotionally. As Pepper explains, "I was created by Softbank Robotics in Tokyo, Japan. I flew all the way to Toronto to work at Humber River Hospital—I love it here."[5] Equipped with sensors and cameras, Pepper has the ability to detect emotions and adapt its own behavior to support the patient. Another example of robot care are the robotic baby seals who were created to comfort the 2012 Japanese tsunami victims. Robot pet therapy has continued, especially in circumstances that prevent much physical contact or situations in which limited caregivers are available.

Robots have long been researched, developed, and used. Economic pressures will continue to spur development. As robots move from functional factory machines to homes, they will become increasingly integrated into our daily lives and, when desired and appropriate, they will be human-like. Robotics, combined with the coming AI explosion, will open up more possibilities for radical affective enhancement.

Datifying Emotions

The collection, manipulation, and use of data is increasingly important in a technological world. Social media companies are prominent among the entities, including government, that track and collect data related to our moods and other aspects of our lives. Usually unknowingly and unintentionally, we tell these companies about our moods through emojis, comments, likes, tags, photos, links, purchases, and social media recommendations. The collected data is then utilized to sell products and guide behavior.

Smart watches, smart phones, and other health wearables monitor sleep, exercise, breathing, skin conductance, and heart rate, all related to how we react to stress-inducing situations. When stress responses are detected, devices measuring these human functions can guide us in the use of calming techniques,

[3] The popular science-fiction romantic film, *Her*, depicts a man having a romantic relationship with Samantha, an artificially intelligent computer. For an example of how complicated scenarios can get, it is revealed at the end of the film that Samantha is having simultaneous romantic relationships with hundreds of human lovers.

[4] Peter Robinson, "Fixed Points in a Changing World," in *Spiritualities, Ethics, and Implications of Human Enhancement and Artificial Intelligence*, ed. Chris Hrynkow (Delaware: Vernon Press, 2019), 227.

[5] Aalia Adam, "Meet Pepper—Canada's First Emotionally Sensitive Robot for Sick Kids," Interview with *Global News* (May 7, 2018). https://globalnews.ca/news/4180025/pepper-Canada-robot/.

such as listening to relaxing music, meditating, or practicing breathing exercises. The Internet of Things (IoT) and the Internet of Bodies (IoB) are changing how we understand, recognize, and make use of human emotions. AI programs that read and analyze facial expression, eye movements, body language, and voice pitch and patterns, can be added to IoT and IoB devices.

Increasingly, AI devices that monitor and interpret affect will be embedded into products. Home entertainment devices will present music, video, or gaming options that fit the consumer's mood. Affectiva, an emotion measurement technology developed in the MIT Media Lab, offers software that analyzes speech, identifying emotional states associated with laughter, voice pitch, and arousal.[6] These datified emotions are used to identify attractive video games, for example.

Refrigerators of the future might suggest food options suited for particular emotional states. A stressful Zoom meeting with colleagues may be identified by the smart refrigerator, which then suggests chocolate milk, or whipped cream and strawberries, or whatever your comfort food might be. Perhaps Siri, Alexa, or Google tells a joke or story to lift your spirits. Combine these personalized databases with humanoid robots, and the possibilities for affective enhancement become clearer.

Brain Stimulation and Other Neuroscience

As discussed in earlier chapters, deep brain stimulation (DBS) can be an effective treatment for intransigent clinical depression, including the depression associated with Parkinson's Disease.[7] By relieving clinical depression, DBS can improve one's affective state to such a degree that it makes life worth living again for some people. It may be that transcranial stimulation (TCS) can also help overcome neurological blocks to feeling happy and content.

We now record and interpret brain wave patterns in support of techniques, such as biofeedback, to help people change brain patterns and, therefore, moods. As mentioned in the chapter, "Cognitive Enhancement and Moral Bioenhancement," neuroscience research is pursued by some military organizations to address fear and traumatic memory formation.[8]

[6] "Affectiva." www.affectiva.com.

[7] Ettore A. Accolla and Xlauda Pollo, "Mood Effects After Deep Brain Stimulation for Parkinson's Disease: An Update," *Frontiers in Neurology* (June 14, 2019). https://doi.org/10.3389/fneur.2019.00617.

[8] M. N. Tennison and J. D. Moreno, "Neuroscience, Ethics, and National Security: The State of the Art," *PLOS* (2012). http://journals.plos.org/plosbiology/article?id=10.1371/journal.pbio.1001289; and R. K. Pitman, et al., "Pilot Study of Secondary Prevention of Post-Traumatic Stress Disorder with Propranolol," *Biological Psychiatry* 51, no. 2 (2002): 189–192.

Technology of Spiritual Enhancement

One of the ways we have sought to find deeper meaning and feel better is through spirituality. Spiritual practices and the benefit they provide are certainly a part of the devotional expressions of all religions. Our understanding of spirituality, however, is broad enough to include spirituality that can occur outside the context of traditional religions.

Psychologist Kenneth I. Pargament and his colleagues have used quantitative and qualitative research studies to identify three key (but not exhaustive) qualities that characterize spiritual experiences: transcendence, ultimacy, and boundlessness. Transcendence is the feeling of "mystery and ineffability." Ultimacy has to do with a sense of meaning deeper than meets the eye. Boundlessness points to timelessness and spacelessness experienced when encountering what is perceived as sacred.[9] Later, Pargament and his colleagues added the spiritual qualities of interconnectedness and spiritual emotions.[10]

Spiritual emotions are emotions accompanying encounters with the sacred. Spiritual practices and experiences can generate spiritual emotions, such as awe, elevation, hope, joy, and gratitude. Sociologist Émile Durkheim's notion of "collective effervescence"[11] describes the affective arousal occasioned by communal religious gatherings. The experience of these emotions may be so intense to be disturbing to some, therefore not desirable. It should be obvious that affective enhancement and spiritual enhancement are not the same, but they do overlap in significant ways. It is not accidental that both are addressed in the same chapter of this textbook.

Pargament's understanding of spiritual experience is similar to the experience reported by mystics in both monotheistic and karmic religions. Mysticism is here understood as a religious experience that is noetic (involves deep knowing), ineffable (cannot be articulated), holy, characterized by a positive affect, perceived as timeless and spaceless, and characterized by a perception of unity in all things. This understanding of mysticism is distinct from paranormal or psychic experiences, as in extrasensory perception and clairvoyance.[12]

Since spirituality, according to Pargament, is basically a search for the sacred, he has found that people can discover the sacred in almost anything. Places where people discover the sacred and so experience spirituality include

[9] K. I. Pargament, D. Oman, J. Pomerleau, and A. Mahoney, "Some Contributions of a Psychological Approach to the Study of the Sacred," *Religion* 47, no. 4 (2017): 723–724.

[10] J. W. Lomax, J. J. Kripal, and K. I. Pargament, "Perspectives on Sacred Moments in Psychotherapy," *The American Journal of Psychiatry* 168, no. 1 (2011): 12–18; and A. Mahoney, K. I. Pargament, B. Cole, T. Jewell, and R. Phillips, "A Higher Purpose: The Sanctification of Strivings in a Community Sample," *The International Journal for the Psychology of Religion* 15, no. 3 (2005): 406.

[11] Émile Durkheim, *The Elementary Forms of the Religious Life* (New York: Free Press, 1912/1965).

[12] These marks of traditional mysticism were identified by philosopher W. T. Stace. See Calvin Mercer and Thomas Durham, "Religious Mysticism and Gender Orientation," *Journal for the Scientific Study of Religion* 38, no. 1 (1999): 175–82.

relationships, objects, and activities, such as found in sport, music, art, family, romantic partners, writing, gardening, and walking. This may seem like an anything-goes type of spirituality, but that is not the case.

For the sacred to be authentically discovered, Pargament and his colleagues identified not only key qualities of spiritual experiences, but also six implications for everyday living that result from the discovery of the sacred. These implications include investment of resources, such as money and time, and the generation of spiritual emotions, as two of the six implications.[13] Pargament has provided a framework that we think can assist in assessing traditional and radical enhancement methods that may help us achieve more, and possibly more dramatic, spiritual experiences.

To illustrate this broad understanding of what counts as spirituality, consider this quote from one of your authors, Professor Trothen, from her book about sports, enhancements, and spirituality.

> Mainstream religions are commonly considered authentic, while the religious dimension of fandom has often has been considered fake or dismissed as derivative. … [religious studies scholar Jennifer] Porter makes a convincing case for the "authenticity" and validity of "pop-culture inspired spiritualities:"[14] fan communities, she proposes, "are, or at least can be, a place that embodies a person's and/or a community's expression of the essence of all meaning: what it means to be human, to be in community, to be in space and time, to be moral or immoral, to be finite or eternal, to simply be…."[15] The question of meaning need not be explicitly pursued, but can be lived or experienced. The implicit quality of spirituality inspired by pop culture, religions, or both, resists hard-and-fast definitions, just as they open up a multiplicity of ways of experiencing the sacred.[16]

If we begin with the premise that the sacred can be discovered in many places, then the potential for enhancing spiritual experiences is expanded. Our discussion is deliberately wide-ranging. We will next present some possible ways to radically enhance spirituality. We have provided some evaluative comments about authentic spiritual experiences, and leave up to you a consideration of whether radical enhancements might help us to generate authentic spiritual experiences.

How We Make Ourselves Spiritually Better

With Pargament's work as a foundation, we now consider the variety of ways to achieve experiences with spiritual qualities like transcendence, ultimacy,

[13] Pargament, et al., "Some Contributions," 734.

[14] Jennifer Porter, "Implicit Religion in Popular Culture: The Religious Dimensions of Fan Communities," *Implicit Religion* 12, no. 3 (2009): 272.

[15] Porter, "Implicit," 275.

[16] Tracy J. Trothen, *Spirituality, Sport, and Doping: More than Just a Game* (SpringerBriefs Sport and Religion Series. Switzerland: Springer International Publishing, 2018), 3.

boundlessness, interconnectedness, and spiritual emotions. Since we are understanding spirituality in a way that does not confine it to traditional religious structures, our discussion is potentially applicable to everyone, whether they identify with one of the traditional religions or not. As we did above in the affective enhancement section, we begin with traditional methods of spiritual formation and then consider some of the radical therapies and technologies.

While the spiritual enhancements discussed below may add much to spirituality, it is hard to imagine that these enhancements will be enough on their own to lead to greater spiritual well-being or meaningful experiences. But they may, and time will tell as the religions assess and utilize some of the increasing array of spiritual enhancement intervention options.

Traditional Spiritual Disciples and Practices

Professor Trothen has a Muslim neighbor who spiritually enhances themself by praying five times a day facing Mecca, a traditional Muslim practice. The neighbor's prayer discipline has affective benefits as well, seeming to help them stay happy and content most of the time. While completing this textbook, Professor Trothen was in Kuujuaq, an Innuit village in Nunavik, Canada, on a beautiful edge of the tree line. Many Innuits find their spirituality enhanced by spending time on the land, appreciating the rivers, tundra, taiga, and rocks that shape the small mountains of the region. A blend of traditional Innuit spirituality and Christianity shapes the spiritual practices of many of these far north Indigenous communities.

These two examples come from traditional religion and are joined by a host of other long-standing spiritual practices that include spiritual direction, guidance, contemplative practice, and mindfulness meditation. While these practices do not require advanced technology, technological devices can be used to facilitate the practices, such as tablets with software that leads one through meditative practices.

Technology can also help us to understand the effects of traditional spiritual practices. For example, scanning technology correlates brain activity with traditional spiritual practices. Brain scans show that religiously based meditative practices produce effects also achieved in "flow states," a term coined by psychologist Mihaly Csikszenthmihalyi. Flow states are characterized by total absorption in the experience, the sense that all life is connected, a strong sense of self, and the loss of individual ego. Flow states are not restricted to formal religious practices or to self-described religious followers.[17]

[17] Csikszenthmihalyi describes eight elements of flow states: clarity of goals and immediate feedback; a high level of concentration; a close match between one's perceived skills and the challenge; a feeling of control; effortlessness; an altered perception of time; the melting together of action and consciousness; and the experience of the autotelic quality of the sport. For more explanation and analysis, see Susan Jackson and Mihaly Csikszentmihalyi, *Flow in Sports: The Keys to Optimal Experiences and Performance* (Champaign, IL: Human Kinetics, 1999). For a fan to experience

An athlete in flow is aware of working very hard but also experiences the sense of effortlessness or not working at all. Some athletes describe flow as being on automatic pilot or in the zone. Flow can be experienced with many activities, including art, writing, and listening to or playing music. Although flow requires total concentration and emotional centeredness, flow cannot be induced, at least thus far. But the induction of flow experiences could be a future spiritual enhancement. Technology allowing for the identification and study of flow states in traditional spiritual practices is a nice segue into using technology itself to generate spiritual experiences, apart from traditional religious practices.

Hallucinogenic Agents

An old Beetles' song is "Lucy in the Sky with Diamonds," a not so subtle reference to lysergic acid diethylamide (LSD), one of the hallucinogenic drugs of choice in the counter-culture hippie era of the 1960s. What does that have to do with human spiritual enhancement in our century? Christian theologian and ethicist Ron Cole-Turner, introduced in the chapter, "Transhumanism, the Posthuman, and the Religions," distinguished the category of spiritual enhancement. Cole-Turner's focus is on the use of hallucinogenic drugs to open-up an enhanced spiritual awareness and experience.

Hallucinogenic agents have long been used in a number of religious traditions.[18] The ancient Hindu Vedic scriptures from India speak of *soma*, described as a plant with no roots, no leaves, no fruit, no seeds, but with a white stem, a red cap, and a juice that was golden. "We have drunk soma and become immortal; we have attained the light, the Gods discovered."[19] The Vedic and Aryan warrior God Indra liked his *soma*, and his strength increased under its intoxication.

Use, in a limited fashion, of ayahuasca and mescaline (derived from peyote, a cactus) is allowed in the United States under the First Amendment's free exercise of religion clause. Ayahuasca is used in the syncretic Christian churches União do Vegetal and the Santo Daime. The Native American Church has an exemption for the sacramental use of peyote. Health Canada has granted exemptions to two Montreal religious groups that stem from the Brazilian religion Santo Daime, the Eclectic Centre for the Universal Flowing Light, also known as Céu do Montréal, and the Beneficient Spiritist Center União do Vegetal. The exemptions allowed for the import and serving of ayahuasca and chacruna, both of which have hallucinogenic properties, to its members.

flow, they need to identify strongly with the athlete(s) For example, the fan must be fully convinced that the athlete's abilities make the athletic challenge possible but not easily possible.

[18] Robert Jesse, "Entheogens: A Brief History of Their Spiritual Use," *Tricycle: The Buddhist Review* 6, no. 1 (Fall 1996).

[19] *Rigveda* 8.48.3, in Ralph T. H. Griffith, trans, *The Hymns of the Rig Veda* (Benares, India: E. J. Lazarus and Company, 1896).

Followers of these religions believe the ingestion of these plants in tea can lead one to meet the divine.

"Entheogen," from the Greek, literally means "full of god" (*entheos*) and "to come into being" (*genesthai*). The term is used to refer to the use of hallucinogenic agents in religious contexts. Academic research on entheogens goes back at least to the research of Harvard professors, Drs. Timothy Leary and Richard Alpert. Amid media fanfare, both were kicked out of Harvard for their ethically problematic research program.

Dr. Alpert found a guru in India who gave Alpert the name Ram Dass, the name by which he is most well-known. Following his return from India, Ram Dass became an important teacher, bringing Hinduism to the counterculture west. As Dr. Alpert, he helped conduct the famous—or infamous as the case may be—Marsh Chapel experiment, performed in the chapel at Boston University on Good Friday, 1962. Seminary students received either psilocybin, derived from certain mushroom species, or a placebo. It is generally agreed there was a positive correlation between the hallucinogenic agent and deeply spiritual, i.e., mystical, states. Many of the subjects reported having profound, and even life-changing, spiritual experiences, including a strong sense of connection with all life. That said, the experiment was not widely reproduced and contained ethics problems and design flaws. The research was very controversial and not supported by the university.

Research on hallucinogenic agents died out for several decades in the United States due in large part to laws intended to halt recreational use of these agents. With research exemptions, scholars have recently revived this line of research in Europe and the United States in controlled medical settings, with a focus on psilocybin. The research shows that psilocybin can be safely and reliably correlated with mystical experiences in healthy volunteers.

Researchers are careful to say psilocybin "occasions" the mystical experience but does not necessarily "cause" it. This means psilocybin at the least creates conditions for possible mystical experiences. And, based on a research project that surveyed thousands of people, it seems that experiences of personal encounters with God can occur for previously self-identified atheists (more than two-thirds of whom stopped calling themselves atheists after their encounter). Moreover, regardless of whether the spiritual experience was spontaneous or occurred while taking a psychedelic, a majority of respondents who reported such God encounters also reported lasting positive changes to their mental and emotional health.[20] Research on the relationship between psychedelics and spirituality continues, at the time of the writing of this book, at the Johns Hopkins School of Medicine in the United States.

[20] Roland R. Griffiths, Ethan S. Hurwitz, Alan K. Davis, Matthew W. Johnson, Robert Jesse, "Survey of Subjective 'God Encounter Experiences': Comparisons Among Naturally Occurring Experiences and Those Occasioned by the Classic Psychedelics Psilocybin, LSD, Ayahuasca, or DMT," *PLOS ONE* (April 23, 2019). https://doi.org/10.1371/journal.pone.0214377.

Your authors know people who have participated in recent research studies on the use of psilocybin with regard to spiritual experiences. The experiences occasioned by psilocybin are indeed profound and, apparently, the evidence thus far is that the effects are lasting, at least for several years and possibly for one's lifetime. Here are some first-hand descriptions of these psilocybin experiences:

> Feeling of a non-self, of self held/suspended in an almost tactile field of light.
>
> The utter joy and freedom of letting go—without anxiety—without direction—beyond ego self.
>
> The sense that all is One, that I experienced the essence of the Universe and the knowing that God asks nothing of us except to receive love.
>
> The experience of death, which initially was very uncomfortable, followed by absolute peace and being in the presence of God. It was so awesome to be with God that words can't describe the experience.[21]

This research raises questions about what, if any, appropriate role there might be for the use of such pharmacological agents in religious or spiritual practice and ritual. Some people find this suggestion troubling and see psilocybin as nothing more than illicit drug use or a form of spiritual cheating, raising the question of what constitutes authentic spirituality. Others, following a proactionary stance toward spiritual enhancement, are open to the possible beneficial consequences of psilocybin induced experiences and want more research. This debate is sure to continue and perhaps intensify.

Brain "Spirit Tech"

In this section, we present a host of developing technological innovations, sometimes called "spirit tech," that allegedly provide authentic spiritual experiences. These technologically-assisted spirituality biohacking tools may have some of the same properties as hallucinogenic agents, but with a more precise focus on relevant brain centers and, therefore, possibly with more control of the outcomes.

Neurofeedback, a type of biofeedback, may have benefits in treating Attention Deficit Hyperactivity Disorder (ADHD) and other medical conditions. With sensors on the scalp, the patient obtains real-time feedback as they learn to self-regulate brain activity. Neurofeedback is now being examined as a way to enhance meditation and prayer practices. As mentioned earlier, transcranial magnetic stimulation (TMS) is a relatively safe, non-invasive procedure sometimes used to treat major depression. Via an electromagnetic coil near the forehead, the part of the brain associated with mood and depression is stimulated. TMS may hold possibilities for activating the so-called "God

[21] Ronald Cole-Turner, "Spiritual Enhancement," in *Religion and Transhumanism: The Unknown Future of Human Enhancement,* eds. Calvin Mercer and Tracy J. Trothen (Westport, CT: Praeger, 2015), 369–84.

spots" of the brain, which are the areas of the brain thought to be responsible for spiritual experiences.

The connection between neuroscience and spirituality has intrigued researchers for years. Functional Magnetic Resonance Imaging scans (fMRIs) have shown the effects spiritual practices, including centering prayer, mindfulness practices, meditation, guided imagery, and music, have on the brain.[22] Digital technology companies are capitalizing, for example, on the knowledge that music can make one run faster by stimulating the brain. Spotify uses phone sensors to measure a runner's tempo and chooses music to help the runner keep or improve their pace.

A good bit of publicity has surrounded the controversial "God Helmet," which you can order online for several hundred Canadian dollars. Created by a neuroscientist, Stanley Koren, and based on research by the late professor of psychology, Michael Persinger, the helmet is a device with electrodes used to study creativity, religious experiences, and the effects of stimulation of the amygdala and the hippocampus. While the efficacy of this particular item is unproven, the world of Spirit Tech has arrived.

Experimentation is underway with small groups of meditators/worshippers, all of whom are using the same brain activation method in an effort to foster brain-to-brain interaction and collective experience. The serious caveat to all this potential, as we discussed earlier, is that many of these brain stimulation technologies may cause seizures or headaches and may change thought patterns, affecting personal identity in unforeseen ways.

Pixel Spirituality

Other innovations are afoot that have been changing, and may radically enhance, spirituality. Professor Mercer glimpsed these technological possibilities when he discovered that a Christian friend of his was "going to church" on Sunday mornings by going on-line. What made this news striking is that Mercer's friend is 68 years old and all her life has advocated a conservative version of Christianity practiced in a very traditional Baptist church. This friend's willingness to engage in this method of worship, even before the COVID-19 pandemic, indicates a trend toward openness to "pixel spirituality."

We have seen much more virtual worship since the COVID-19 pandemic began. When large gatherings were prohibited and physical distancing measures implemented, people who otherwise may have never explored platforms, such as Zoom, Facebook, Skype, or Microsoft Teams, found themselves conducting worship or participating in worship digitally. Virtual worship has not replaced

[22] S. E. Kobar, et al., "Ability to Gain Control Over One's Own Brain Activity and its Relation to Spiritual Practice: A Multimodal Imaging Study in Frontiers," *Human Neuroscience* (2017). https://doi.org/10.3389/fnhum.2017.00271; and A. M. Schultz and P. E. Carron, "Socratic Meditation and Emotional Self-Regulation: Human Dignity in a Technological Age," *Journal of Interdisciplinary Studies* 25, nos. 1–2 (2013): 137–160.

face-to-face as the preferable option for everyone, and not all were able to join due to internet and technology accessibility, but the virus made this digital avenue a viable option for many. Current virtual worship platforms can be viewed as an interesting technological update to the plethora of TV preachers that flooded the television airways in the 1980s. Virtual worship trends raise questions about community in the religions, a concern we address later in the chapter.

The ability to worship on the internet with like-minded friends is a primitive version of the technological spirituality that is alive and well in virtual worlds. "Second Life" is a popular example of virtual reality that has been around for years. For those largely unfamiliar with virtual worlds, Second Life has been an industry leader in creating a world that exists in cyberspace and is usually accessed through a computer keyboard and the internet. Robert M. Geraci is the author of *Virtually Sacred: Myth and Meaning in "World of Warcraft" and "Second Life"* (we will leave aside "World of Warcraft" in our discussion). Geraci explores how virtual worlds, like Second Life, are "rearranging or replacing religious practice."

Although not a regular visitor, in order to understand this platform for spiritual enhancement, Professor Mercer spent about 15 hours in the fascinating virtual world of Second Life. Here is how it works. You, the "player," sit down at your computer and log into Second Life. In your first visit, you get to pick an avatar, an image that you want to "be" you in the virtual world. Your avatar can be male or female, or androgynous—whatever you like. You can pick any kind of body, any clothing, anything at all. Many people experiment with "being" someone quite different from who they are in real life. Using keyboard strokes, you move your "self" (i.e., avatar) around and communicate. To make a long story short, you design your avatar, buy clothes purchased with currency utilized by Second Life "citizens," visit virtual cities, dance in clubs, have a beer with a friend at a local pub, and, yes, join a church or other faith community.

For better or worse, players in the virtual reality can negotiate their faith identity, worship with their chosen community, engage in spiritual practice, and have religious experiences. As Geraci puts it: "logging in is, for many users, a sacred opportunity to experience what they see as a tiny fraction of the heavenly world to come."[23]

We have the strong sense something is aborning here that is altering our religious landscape and may affect the transhumanist agenda for human enhancement in unforeseen ways. The impact is going to increase in the coming years and in ways that we can only vaguely anticipate. People who are not very mobile, or who are living in the midst of a pandemic, can "get out" into another world and live a whole different life. Virtual world adventurers can join faith communities without necessarily connecting with anyone encountered in

[23] Robert M. Geraci, *Virtually Sacred: Myth and Meaning in "World of Warcraft" and "Second Life"* (Oxford and New York: Oxford University Press, 2014), 200.

their real world. Perhaps virtual worlds will enhance faith and spirituality. Maybe it will not. More likely, digital spiritually will have pluses and minuses.

Religious Issues

The Problem of Suffering

Perhaps it seems obvious that we should work to relieve pain and suffering, whether it is emotional, cognitive, physical, spiritual, or other. We now place the conversation about affective and spiritual enhancements in the context of what has been called the "problem of evil and suffering." In summary, the problem is that in the monotheistic religions God is understood to be all-knowing, all-powerful, and all-loving. Omniscient, omnipotent, and omnibenevolent are the technical theological terms referring to these divine attributes.

In the face of a divine being with these attributes, how can evil, pain, and suffering exist? In other words, why would a good and loving God, who can do anything, allow pain and suffering? If God is all-good, then God would not want people to suffer. If God is all-wise, God can figure a plan to eliminate suffering. If God is all-powerful, God can implement any plan that is conceived.

Theological reflection on this topic has been so extensive through the centuries that the term, theodicy, literally Greek for "God justice," was coined. How can one justify the existence of a supremely wise, powerful, and good deity in the face of such suffering and evil? Many books have been written about suffering and religious belief including one of the most well-known modern ones, *Why Bad Things Happen to Good People*, by Rabbi Harold Kushner.[24]

Process theologians propose that God is in relationship with human beings. Relationships are dynamic and both parties, God and humans, change as a result of being in relationship. God, in this view is not all-powerful, at least not in the way that all-powerful is usually understood. God does not exert power to rescue us but does exert power to love and support us, in this view. Liberation theologians understand God to be in solidarity with those experiencing injustice and understand that the work of justice is inspired and supported by God. These are very quick and simplistic summaries of some complex theological interpretations. Suffice it to say there are diverse views on suffering and God's role in suffering.[25]

The problem of evil and suffering has a very different frame in the karmic religions. Since these religions have no all-knowing, all-powerful, all-loving deity, evil and suffering present no challenge to god as they do in monotheistic

[24] Harold Kushner, *Why Bad Things Happen to Good People* (New York: Anchor Books, 1981).

[25] See, for example, Douglas John Hall, *God and Human Suffering* (Minneapolis: Augsburg Publishing House, 1986) and Traci C. West, *Solidarity and Defiant Spirituality: Africana Lessons on Religion, Racism, and Ending Gender Violence* (New York: New York University Press, 2019).

frames of reference. In general, suffering in the karmic religions is understood as a result of bad karma in previous incarnations.

Christian pro-enhancement advocates propose that God does want to alleviate suffering, and technological advance is a primary means of doing that. Much energy and many resources go into alleviating suffering individually and as a society. Perhaps a pill that radically reduces suffering would be desirable. Maybe, but this is a problematic proposition.

Polio and smallpox vaccines save millions of lives a year, and Prozac reduces the mental anguish endured by depressed patients. It is a good thing that smallpox has been eradicated. And we certainly want to eliminate all emotional and mental pain and disturbance that we can. Not so fast. Here we bump again into the notion of unintended consequences.

Consider a specific example, a case where a seemingly obvious response turns out to be problematic. In this example, a pharmaceutical eliminating aggressive behavior is developed. By decreasing impulsive aggressive behavior in people convicted of violent crimes, perhaps that would increase community safety. But, should aggressive behavior always be avoided? At times, controlled aggression might be appropriate in the workplace and, unless you are a pacifist, in defending one's country. Consider acts of heroism in which people save someone by violently pushing them away from an oncoming subway train. Some really aggressive behavior results in good outcomes and can be judged virtuous.

Pain and suffering serve beneficial purposes, at least sometimes. At a very basic level, pain in my chest alerts me to inquire about the need for medical attention. Maybe we need to exercise more so that mild chest discomfort is addressed by better cardiac health. Earlier we introduced the DREAM gene that is linked to the perception of pain. Mice who do not have this gene have greatly reduced sensitivity to pain. It is hoped that more research will yield interventions that block physical pain, while maintaining a level of sensitivity necessary for good health.

Suffering is about more than physical pain. Suffering has a clear downside but also can have an upside. Suffering, such as existential angst, may lead us to connect more with other people in the journey of life. Pain and suffering can prompt wise reflection on life options. While it does not solve the problem of theodicy for the monotheistic religions, we should note that many, and perhaps most, people agree they have learned much from suffering. Although unfortunately not always redemptive, suffering sometimes can be a powerful teacher and build character.

Of all the religions, Buddhism has placed suffering at the center of its theological program. Suffering is the fundamental predicament of life, the basic problem, in this religion. It is typical to use a medical model here. The presenting problem or symptom is suffering, the diagnosis is craving or desire causing the suffering, and the prescription or cure is to eliminate desire, thereby eliminating the suffering. If, and this is a very big if, radical therapies and technologies were to one day eliminate pain and suffering, Buddhism would need to engage fundamental theological work to reframe how it understands human suffering. Even now, and in the foreseeable future, technology that greatly reduces certain types of suffering merits Buddhist theological reflection. Of course, it is highly unlikely that technology will solve the problems of desire and attachment, but the reduction of suffering may generate questions in Buddhism about the nature of human suffering and our responses to this suffering.

Finally, the psychological argument has been made that "contrast experiences" are necessary. Humans have to experience some measure of unhappiness, or suffering, to know and value happiness. Mountains do not come without valleys is an old saying that holds some psychological and spiritual truth.

The Quick Fix or Effort and Discipline?

Perhaps it is impossible to be truly happy, content, and at peace without the hard inner work of processing experiences, building relationships, addressing our wrongs, and in a healthy way confronting those who have wronged us. Spiritual enhancement through biohacking is not likely to be sufficient in itself, but perhaps the religions will find that it can play a role in the spiritual path.

Taking a pill and logging onto a computer to have a spiritual experience takes effort, but certainly not the kind and amount of effort typically required for that experience. About once a year, Professor Mercer devotes the time and effort required for a weeklong spiritual retreat, often at a Roman Catholic Trappist or a Hindu monastery. Making this happen entails enormous scheduling considerations, packing, hours of driving, financial costs, time and effort in prayer and meditation, and a good bit of talking to people.

With all the current and looming technologies for enhancing both affect and spirituality, the role of effort and discipline becomes a central issue. Conceivably, effort and discipline are essential ingredients in the formation of a spiritual life and a happy life. However, if so, it is not at all clear how much effort and how much discipline are required. The same question can be raised about knowledge required to make spiritual experiences meaningful. The concern is that these technologies are a quick fix that cheats the spiritual seeker from authentic growth and the person who seeks emotional well-being from a greater wholeness.

It would certainly be much quicker and easier to swallow a pill, if that could provide the same emotional and spiritual benefits. On the one hand, it could be that the end-result experience is what matters most or, on the other hand, it could be that the process of getting to that experience is the essential component. Professor Trothen deliberately spends time in nature as part of her spiritual discipline. She finds that the process of being present in nature and being still in the moment is not only calming but helps to increase her attentiveness to ecojustice and the grave importance of figuring out how to make sacrifices, develop policies, and save our planet. It may be that radically enhanced spiritual experiences may help us to deepen our awareness of the interconnection of life more quickly and spur us to address climate change more radically.

At one level, the question of effort is an empirical one—is the nature and extent of spiritual experience or mood produced by biohacking the same as, or of the same value as, that produced by traditional methods of spiritual practice or the hard work of developing emotional health? Measurements for emotional well-being are more advanced than are measurements for spirituality. We have tools for measuring spiritual distress and some for making spiritual assessments. Theoretically, we will also eventually be able to measure spiritual or mystical experiences, but we are certainly a long way from doing that reliably.

Some spiritual assessment scales have been developed that have been assessed as reliable and valid tools, in social scientific terms. But, as some critics have noted, it can be very difficult and unhelpful to reduce spiritual health to a set of Likert scale indicators.[26] At the same time, these assessment tools provide spiritual health professionals with a way to begin building a spiritual diagnosis and treatment plan. The downside is in turning to a social scientific tool as reflective of a person's Gestalt spiritual health.

For example, the Spiritual Distress Assessment Tool (SDAT) includes five items assessing unmet spiritual needs, associated with poorer health outcomes, in elderly hospitalized patients. The SDAT is helpful in identifying people who would benefit from spiritual care.[27] However, the SDAT itself will not tell us the wider, often complex, narrative that gave rise to the unmet spiritual needs. The complexity of individual narratives unfolds only over time. To illustrate, for a cancer patient refusing chemotherapy, the SDAT could indicate unmet spiritual needs and what a spiritual effective treatment plan could look like. But the tool itself will not indicate the exact causes of the spiritual distress. To figure out the causes, one must be a skilled expert and listen to the person's story, responding in ways that illicit more and preserve safety. Professor Trothen is certified as a spiritual health practitioner and supervisor by the Canadian Association of Spiritual Care. That certification required a process involving years of intense training and practice under the supervision of trained professionals. In other words, human touch, effort, and discipline are still very much needed in the practice of spiritual health.

The question of effort can be framed theologically. It might be that long term discipline, education, and mentoring are potentially salvific, but it may also be that pills and virtual reality are potential means of grace, that is, fresh new ways that God is making available alternative spiritual paths. Such enhancements, in other words, could be more extreme versions of microphones used in worship and Kindle tablets to read the Bible, or other low-tech devices

[26] Emily K. Trancik, "Lost in Translation: Spiritual Assessment and the Religious Tradition," *Christian Bioethics* 19, no. 3 (2013): 282–298.

[27] Stefanie Monod, Estelle Martin, Brenda Spencer, Etienne Rochat, and Christophe Büla, "Validation of the Spiritual Distress Assessment Tool in Older Hospitalized Patients," *BMC Geriatrics* 12, no. 13 (2012). http://www.biomedcentral.com/1471-2318/12/13.

no one is questioning. Recalling the earlier discussion of suffering, it seems that if there is a quick path to a deeper spiritual experience that is life-giving and profoundly inspiring, then it would be good to take that short path if suffering is relieved and overall well-being improved.

Finally, the answer to whether effort is required for spiritual and emotional health may be yes and no, depending on the affective or spiritual enhancement at hand and the particular individual's needs. One might conclude virtual reality is an appropriate technological support for spiritual formation and happiness but ingesting illegal hallucinogenic agents is not. If so, then we need to be clear about the criteria for drawing these different conclusions. Also, using Second Life for spiritual enhancement and emotional support may be acceptable if it is used in certain ways, perhaps as an adjunct to more traditional forms of spiritual formation and psychotherapy.

We do not presume to have the answers, but these kinds of questions will confront the religions in the brave new technological world increasingly a part of our lives. Commentators are already beginning to weigh in on the advisability of spirit tech. Just to give an example from the conservative side of the Christian theological spectrum, the marketing for one book states that the author "is a Christian who loves God and is attached to his iPhone."[28] The author grants that technology can be a tool to know God, but argues that caution is in order.

The Importance of Community for Spiritual Growth

All religions, some more than others, require and foster community. In Christianity, *koinonia* (Greek, communion or fellowship) is the word often used to describe the early church community with its members, at least ideally, thoroughly committed to each other and having "all things in common."[29] Similar expressions of the importance of community are found in other religions.

As religions embrace, to one degree or another, spiritual enhancement, the faithful will likely insist that the technology be used in such a way that it does not shortchange genuine community. This insistence may well be based in the human need for social interaction, as well as in theology. Ideally, technology supports and enhances the development of community that empowers individual believers in the faith community.

If Second Life is used in spiritual practice, it should not be commandeered to substitute for engagement in healthy social interaction. If entheogens are to be used, they certainly should be used legally, with proper guidance, and in contexts conducive for spiritual attainment. In fact, the research on such agents

[28] Craig Detweiler, *iGods: How Technology Shapes Our Spiritual and Social Lives and Selfies: Searching for the Image of God in a Digital Age* (Ada, MI: Brazos Press, 2013). See also his book, *Selfies: Searching for the Image of God in a Digital Age* (Grand Rapids: Baker, 2018).

[29] In the New Testament, e.g., see Acts 2:42–47 and 4:32–37.

is clear that set and setting are critical to the experience. That is why recreational use of LSD yields a very different experiential outcome than ingesting such an agent in the context of a spiritual community, long-standing ritual, commitment to spiritual growth, and competent spiritual guides.

Similarly, affective well-being cannot be understood outside of community and relationships. Research has shown that happiness is associated with altruistic behavior, under the condition that one does not self-sacrifice to the extent that one loses or damages oneself.[30] Religions have known this for centuries. Feminist theologians and religious studies scholars have also recognized the dangers of the excessive tendency to self-sacrifice that has been encouraged within marginalized groups. Charitable acts are an important expression and command in Islam. Judaism upholds the preservation of life, and not just the preservation of our own lives but the lives of others, as the highest mitzvah. Daoism and the other karmic religions see altruistic acts as creating good karma. Religions do not mandate good acts just to make followers feel better, but they do so as a faith conviction that enhances one's individual well-being, as well as that of the community, world, and one's relationship with the transcendent.

Beyond the implications of spiritual enhancement for existing religions, we have seen at least one spiritual community emerge that is explicitly organized around superlongevity. We introduce the "Church of Perpetual Life" as an example of a possible new religious movement. The founder of the church is Bill Faloon, director and co-founder of Life Extension Foundation,[31] a consumer advocacy, research, magazine, and supplement store group that is favorable to transhumanism. The Church of Perpetual Life website[32] proclaims that

> Our Mission is to assist all people in the radical extension of healthy human life, and to provide fellowship for longevity enthusiasts through regular, holiday and memorial services.

The website goes on to provide eight purposes of the church, which align the church with the transhumanist agenda in general. The "Welcome" message on the website reads:

> Perpetual Life is the only science/faith based church in the world. We are not a bible based church & although we are not a Christian Church, many of our members are Christian & Jewish. We also have members that are Buddhist,

[30] Stephen G. Post, "Altruism, Happiness, and Health: It's Good to Be Good," *International Journal of Behavioral Medicine* 12, no. 2 (2005): 66–77. https://doi.org/10.1207/s15327558ijbm1202_4.

[31] www.lifeextension.com.

[32] https://www.churchofperpetuallife.org/. Here is a somewhat critical documentary of the Church and one which addresses the relationship between the Church and the Life Extension Foundation. "Worshipping Immortality at the Church of Perpetual Life" (April 11, 2016). https://www.youtube.com/watch?v=EvaC67CeBDA.

Humanist, Atheist, and Hindu. What brings us together as a Family is our Faith in Physical Immortality. Humanity is constantly overcoming obstacles that at first appear impossible, and once overcome, a new era dawns and humanity is elevated. And so it is with our belief in living Unlimited Life Spans. Humanity is on the brink of a new Era where your physical health becomes Optimal and death becomes Optional.

Appropriation of Spiritual Disciplines for Personal Growth

Mindfulness meditation has been appropriated and removed from the context of Buddhism and used as a way to enhance relaxation, reduce stress, and focus attention. The removal of mindfulness from the Buddhist context into an individualistic, secular context, has been criticized as misrepresenting Buddhist mindfulness and losing the greater spiritual and socially transformative meaning of the practice.[33] The Western focus on individual mindfulness practices turns these practices into something disconnected from the traditions in which they originated. While mindfulness practices can contribute to affective and spiritual enhancement, significant aspects of spiritual wellness, at least, might be lost through appropriation into a secular context. Such appropriation can also be seen as contributing to the marginalization and misunderstanding of the originating religious groups.

The above analysis of mindfulness can also be applied to hatha yoga, which has a rich spiritual context in Hinduism and is often placed in a physical health and exercise context in the west. Indigenous spiritual practices have been appropriated in the same way. Sweat lodges and spirit walks are sometimes used with good intentions to enhance wellness but without careful respect for the originating people and spirituality. Connection and meaning are subverted by thoughtless appropriation. As a result, the full meaning of the spiritual practice is compromised, as well as the relationship with Indigenous people or, in the case of mindfulness practices, Buddhists.

These religious and ethical implications need to be carefully considered when we use spiritual practices in a quest to improve our spiritual and emotional health. Part of that health is, indeed, our relationships to the followers of these traditions and to the integrity of religious practices that, when practiced respectfully, enhance the well-being of not just the individual but of all people, creatures, and the earth.

Can Robots Do the Jobs?

We may see an increasing turn to robots to help provide, or to actually provide, spiritual direction and emotional support. At present, in health care facilities

[33] Maria Ishikawa, "Mindfulness in Western Contexts Perpetuates Oppressive Realities for Minority Cultures: The Consequences of Cultural Appropriation," *Simon Fraser University Educational Review* 11, no. 1 (2018). https://doi.org/10.21810/sfuer.v11i1.757.

for example, while robots serve primarily to fill healthcare "staffing gaps,"[34] they are not capable of providing in-depth or complex emotional support and addressing ontological issues associated with spiritual and existential distress. When we grapple with a sense of our value and meaning as we experience physical losses, or when we struggle to accept our mortality, a robot that uses stock phrases is not going to be sufficient.

We do not minimize the value of robots doing physical labor. Today's clunky robot will evolve dramatically in the coming years with increasing cooperation between the field of robotics and rapidly advancing AI capability. It is theoretical now, but conceivably artificially intelligent robots of the future could perform better than today's trained psychotherapist. The AI robot would have immediately available all the theories and techniques of psychotherapy, every case study available, facial and body movement recognition, and, in addition, would never fall asleep during the therapy session.

With the catchy main title, "Your Robot Therapist Will See You Now," this summary paragraph captures some of the possibilities:

> Research in embodied artificial intelligence (AI) has increasing clinical relevance for therapeutic applications in mental health services. With innovations ranging from "virtual psychotherapists" to social robots in dementia care and autism disorder, to robots for sexual disorders, artificially intelligent virtual and robotic agents are increasingly taking on high-level therapeutic interventions that used to be offered exclusively by highly trained, skilled health professionals.

The article goes on to qualify the hopes for AI therapists, noting that "In order to enable responsible clinical implementation, ethical and social implications of the increasing use of embodied AI in mental health need to be identified and addressed."[35]

The exact same kind of conversation can be applied to religious leaders and the services they deliver. We seem to be in the beginning stages of a time when the conversation about many professional jobs has to do with how to effectively integrate AI and robots into the job description, along with the human professional. The time may very well come, however, when full expendability is the question. Perhaps the day will come, for example, when the Roman Catholic Church has a technological way to address its shortage of priests. Or, it may be that AI does not provide the same sense of connection, depth, and competence that a trained, real person can offer. We are seeing rapid changes in AI. The future may bring much promise and, possibly, some limits.

[34] Corinne Purtill, "The Robot Will Help You Now: How They Could Fill the Staffing Gaps in the Eldercare Industry," *TIME Magazine* (November 4, 2019).

[35] Amelia Fiske, Peter Henningsen, and Alena Buyx, "Your Therapist Will See You Now: Ethical Implications of Embodied Artificial Intelligence in Psychiatry, Psychology, and Psychotherapy," *Journal of Medical Internet Research* 21, no. 5 (May 2019). https://www.ncbi.nlm.nih.gov/pmc/articles/PMC6532335/. https://doi.org/10.2196/13216.

Ethical Issues

Affect and Spirituality: What Exactly Are We Trying to Enhance?

Regarding emotional well-being, complicating the discussion about happiness are findings regarding what really contributes to our happiness. Engaging in challenges to overcome barriers perceived as a bit tougher than our skill level yields happy moments. If philosophers like Hubert Dreyfuss and Sean Dorrance Kelly are correct about our need for these everyday challenges and "shining moments" filled with spiritual emotions, then the more we create technology that solves these challenges, the harder it will be to be happy.[36] For example, many of us have dishwashers. The argument is that without these dishwashers, it was easier to engage in an everyday challenge—getting all our dishes done—and succeed, thus giving us an opportunity for happiness.

Others argue that altruistic acts towards strangers is the surest way to happiness and generally feeling good emotionally. As we mentioned earlier, in some research, spirituality has been associated with increased compassion and prosocial—including altruistic—behavior toward strangers.[37] Maybe, if we want to feel better emotionally, it is more effective to work on enhancing spirituality or morality, including kindness.

As with moral enhancement and cognitive enhancement, what it means to feel better emotionally and spiritually is contextual. From a virtue ethics perspective, different emotions and spiritual experiences may be more or less desirable, depending on what situation we are in, who we are, what we want to do, and who we want to become. These desires are influenced by the sources of authority that shape our values and our identities.

Being different and having varying views on what makes us better do not mean we should act as though we are disconnected from everyone else. Different emotions, for example, are more or less desirable depending on what social messages we receive, our values, and who we want to be and become. It is important that we take a careful look at differences, asking what informs the differences. We learn from diverse perspectives and, as we have emphasized, a religiously informed lens recognizes the interconnectedness of all life.

The Therapy—Enhancement Continuum: What It Means to Make Us Better

The therapy or enhancement debate continues to be relevant in this chapter. What it means to be human prompts questions about how we are human and how we change ourselves in acceptable and good ways. For both affect and spirituality, although it is not fully clear where the line is crossed between an

[36] Herbert Dreyfuss and Sean Dorrance Kelly, *All Things Shining—Reading the Western Classics to Find Meaning in a Secular Age* (New York: Free Press, 2011).

[37] Laura Rose Saslow, et al., "The Social Significance of Spirituality: New Perspectives on the Compassion–Altruism Relationship," *Psychology of Religion and Spirituality* 5 (2013): 201–18.

acceptable and unacceptable intervention, we have identified the interconnection of life as one touchstone dividing point applicable to all religions. An enhancement helping us honor the relatedness of all life and make better (almost?) everyone's lives and the earth is likely to be embraced by most religions. But we still have the challenge of figuring out what counts as appropriately better.

Consider drugs that enhance mood. The legalization of marijuana has been a hot topic in many countries. The trend, although with plenty of starts and stops and debate along the way, in Asia and the West has been in the direction of greater use and a loosening of restrictions.[38] Medical pot is therapeutic when it relieves pain in a cancer patient undergoing chemotherapy or radiation treatments. The THC in marijuana creates the high and not all marijuana derivatives contain THC. CBD (cannabis oil) for pain relief does not produce a high, but the pot brownie edible kind and other forms do. Is getting high enhancing, or not? Possibly, there are certain conditions in which mood elevation from marijuana can be considered enhancing, and other conditions in which such a mood alteration is not so considered. The conversation needs to get complicated again. What it means to make our moods better, and who gets to decide what it means to feel truly better, are critical questions.

Consider drugs that enhance spirituality. Using entheogens could be considered spiritual doping. Conceivably, a difference can be identified between using entheogens (if legal for religious use) and using meditation to deepen spirituality. Likewise, maybe a distinction can be drawn between using something like Buddhist mindfulness meditation and using pharmaceuticals, both of which can decrease stress, anxiety, and fear. Some people contend emphatically there are significant differences, and it is important to be clear and precise about what those differences are. Those differences may be related to the effort and discipline required by spiritual practices and deep intrapersonal reflection, as discussed earlier in this chapter. For some, the reason for rejecting pharmaceuticals is the negative image of street drug use and doping, which is perhaps a more questionable argument.

Some critics have labeled listening to music in sport a form of emotional doping, since fMRIs show neural activity that can improve performance. Many athletes run faster when they listen to certain music. Similarly, the controversial super polyurethane swimsuits, already discussed and dubbed "doping on a hanger," made swimming easier, resulting in many broken records. The use of illegal anabolic steroids is also called doping. The term has strong negative connotations and tends to be interpreted as moving into an unacceptable enhancement side of the continuum. What if we change the term "doping" to "enhancing"? Interestingly, that might alter some people's outlook on the acceptability of using entheogens to evoke spiritual experiences or better moods. Language matters.

[38] E.g., Grace Shao, "Medical Cannabis is Gaining Momentum in Asia," *CNBC* (July 14, 2019). https://www.cnbc.com/2019/07/15/medical-cannabis-is-gaining-momentum-in-asia.html.

Much has to do with the values and beliefs embraced by a religion and how these values and beliefs may be understood in terms of what it means to make us truly better. Modalities such as virtual pixel spirituality and digital worship that allow people to connect with others when they otherwise may not be able to connect, potentially enhance spirituality and in so doing allow us to become more deeply human. For our purposes, the dividing point of what counts as an acceptable or unacceptable enhancement on this continuum rests not on a secular notion of "normal," but rather on a religious understanding of what makes us better, keeping in mind that each religion can define what makes us better in varying ways.

Choice

As with other enhancement categories, many transhumanists frame radical human enhancement as mainly a matter of individual choice. Opting for an affective or spiritual enhancement should be at one's personal discretion. However, as we know and as the religions emphasize, decisions are not only about the decision maker, although it can be difficult to see how choices affect others and why the person choosing should care.

A burgeoning boost to spirituality, with affective enhancement implications, are pixel spiritualities such as virtual worship and avatar programs. Pixel spiritual practices are found in various religious traditions. Buddhism, for example, is alive and well in Second Life, and a recent study by Gregory Price Grieve provides an analysis and critique that can apply to virtual reality in any religion.[39] Grieve contends that online religious practice can be trivial and even harmful if these online experiences distract from real-life experiences. Virtual reality has even been critiqued as a haven for people who cannot cope in real life. Grieve's research supported his theory that personal relationships were the main deficits in people's lives leading them to Second Life.

That said, pixels on a screen can also open the door to a fascinating, sophisticated, and, according to Grieve, authentic spirituality that can clearly be meaningful to participants. There is always the question of what "real" and "authentic" mean. We have to decide if physicality and face-to-face interaction are essential components of religion. If so, then Second Life and other virtual world spiritualities fail the test.

When religious followers choose pixel spirituality experiences over face-to-face spiritual encounters, they affect not only themselves but the flesh and blood people with whom they worship, study holy text, and share food. During the COVID-19 pandemic, many religious communities figured out creative, online ways to be together. Some members found these methods unsatisfying largely because of the physical limitations on interactions. Others were very

[39] Gregory Price Grieve, *Cyber Zen: Imagining Authentic Buddhist Identity, Community, and Practices in the Virtual World of Second Life* (London: Routledge, 2016).

grateful for a virtual option and enjoyed it. Virtual options may enhance relationships and fill gaps, providing new ways of relating and connecting.

In his study, Grieve argues that a cybersocial being emerges from the feedback between the user and their avatar. So, players ("residents" is the term used in Second Life) are cyborgs, that is, flesh and blood people on the computer keyboard and avatars in the virtual world. The Second Life religious groups meet in a place, built by residents, that serves as a liminal (a space between worlds) alternative to real-life work and home.

Choice raises the issue of consent. Consent is problematic, for example, with clinically depressed or anxious people, especially when some affective enhancements are commercially available to anyone, without professional supervision. Making decisions well when we are not thinking as we would without depression or anxiety (and these are only two possible examples) is complicated. Possible risks of the therapy or technology only adds to the complexity of consent.

It is complicated to make an informed and healthy choice for anything we have not yet experienced, even if the effects are not permanent, if it is not passed to progeny, and if the enhancement is safe. These are big "ifs," which point to the value of standard testing of the intervention, the functioning of regulatory bodies, and the delivery of the intervention under professional supervision. Assuming an enhancement meets the criteria of all three "ifs," still the consumer cannot know what a new induced emotional or spiritual state will feel like for sure, until they are in it. Hallucinogenic agents are excellent examples of what we are talking about. Proper testing, certification by reputable agencies, and professional delivery certainly help us make good choices about powerful enhancements but will not tell us everything.

Justice

Choices about enhancements, by individuals and society, if ethically grounded, will work to promote justice or, at the very least, will not amplify injustice. The danger of amplifying existing systemic power imbalances is one example of a serious potential injustice that runs as a theme through these chapters and is a major consideration of a precautionary approach.

Amplification of unjust power structures, on the front end, comes when enhancements are researched, designed, and funded largely by people with enough social power and access to university education and jobs at well-funded pharmaceutical and technological labs. On the back end, the interventions are usually available primarily, or only, to the wealthy and politically connected with easy access.

Consider the example of pixel spirituality access during the COVID-19 pandemic. A very good friend of Professor Trothen's is a religious leader in a rural Christian pastoral charge that includes two churches. She recorded and posted worship services on social media for her congregations, but she was aware that not all of her parishioners had the technology, internet access, and/or technological know-how and confidence to access these worship services.

Other area religious leaders, of different faith traditions, did not have the resources and/or ability to record and post worship services. Rural communities and poor communities may not have access to sufficiently high band-width internet to utilize the newest technology, and they may not have the education and knowledge base to utilize it even if they have access. Plenty of families cannot afford a computer. These access issues are justice issues, and socio-economic status tends to align with racial bias and privilege. All of these justice issues related to pixel spirituality can be applied to the difficulty of lower socio-economic classes receiving services that contribute to emotional and spiritual well-being.

The religions have a strong commitment to justice. Earlier in this chapter, we addressed the appropriation of traditional spiritual techniques (e.g., mindfulness meditation, sweat lodges, hatha yoga) for modern, secular use. There are additional ethical issues not discussed in these pages, such as profit-making from technologies that utilize personal data of consumers, arguably not always with their explicit and well-informed consent. All of these issues are justice concerns. The costs and benefits must always be weighed, hopefully with the aim of achieving a just outcome.

Questions for Discussion

1. What does it mean to enhance happiness? Does your answer change at all if you deliberately engage a religious perspective?
2. Is more happiness a mood (affective) enhancement? Can you think of any contexts in which happiness would not be an enhancement?
3. How might different religions define happiness?
4. What other emotions, beyond happiness, might be good to change? From a religious perspective, do you think there should be any limits on enhancing these emotions? Why or why not?
5. Can you imagine affective enhancement scenarios that would be very positive? Very frightening?
6. Do you think sex robots have any legitimate role to play in making us better?
7. How do you see the relationship between radical affective enhancement and radical spiritual enhancement? What is the difference between the two?
8. What do you think about using entheogens for spiritual enhancement? How do your values relate to your response to this question?
9. How do you think using Second Life to attend a Bible teaching session in virtual reality might differ from attending Sunday School taught with PowerPoint?
10. How do you see social justice relating to the use of affective and spiritual enhancements? Can you think of any ways such enhancements might help to address social injustice? Reflect on how social justice issues may be exacerbated by particular enhancement technologies discussed in this chapter.

PART III

Special Topics: Going Beyond the Edge

CHAPTER 8

Cryonics: Buried, Burned, or … Frozen

Technology

A Stopgap Measure

As we have seen, researchers are pursuing a number of paths to superlongevity, including biological, genetic, and tissue engineering. Superlongevity is the central long-term goal in the radical physical enhancement category. To the degree that achieving the other four categories of radical enhancement (i.e., cognitive, affective, moral, and spiritual) requires a biological body of some kind, then extending that biological body long into the future is critical, in order to partake of other advances in the various categories of enhancement. Those who die of diseases now will miss out on future cures, superlongevity breakthroughs, and other enhancements. Longevity enthusiasts need a stop-gap program, some way to bridge the gap between death and a future time when therapies and technologies are more advanced. Cryonics may be one bridge.

If, for example, someone dies of a particular type of cancer, cryonics—a Greek term, meaning "icy cold"—is a process that preserves the body in a way that it can be theoretically revived in the future when the cure for that particular cancer is available. Likewise, cryonics offers the possibility of accessing future enhancement therapies and technologies.

Alcor Life Extension Foundation is the largest and most well-known cryonics organization and will serve as our primary example of a cryonics organization and of the industry's main policies and messaging.[1] American competitors are

[1] Alcor Life Extension Foundation. http://www.alcor.org. A helpful Alcor publication is Aschwin de Wolf, Brian Wowk, and Alcor Staff, eds., *Preserving Minds, Saving Lives: The Best Cryonics Writings of the Alcor Life Extension Foundation* (Scottsdale, AZ: Alcor Life Extension Foundation, 2012).

C. Mercer, T. J. Trothen, *Religion and the Technological Future*,
https://doi.org/10.1007/978-3-030-62359-3_8

the Cryonics Institute in Michigan[2] and Oregon Cryonics in Oregon.[3] KrioRus is a non-governmental cryonics organization in Russia, the only one in Eurasia with its own storage facility.[4]

Alcor, located in Scottsdale, Arizona, arguably leads the way in cryonics research, organization, and membership. Max More, whose PhD is in philosophy of mind, ethics, and personal identity, has been an important leader in cryonics and served for years as Alcor's CEO. Currently Alcor Ambassador and President Emeritus, More is a leader in the transhumanist movement, having co-founded the Extropy Institute, an organization that helped create the transhumanist movement.[5]

Cryonics is based on three premises: (1) Life can be stopped and restarted if its basic structure is preserved. In support of this premise, Alcor gives the following suspended animation examples: human embryos routinely preserved for years; adult hypothermia victims surviving up to one hour without lung, heart, and brain activity; large animals surviving three hours of cardiac arrest near 0 degrees Celsius (32 degrees Fahrenheit); and deep cooling used during neurosurgery when the heart must be stopped. (2) Extremely low temperatures and chemical agents can preserve the structure of organs, including the brain. (3) Molecular nanotechnology, discussed in the chapter, "Existing and Possible Technologies," is one possible future technology that can facilitate tissue repair and regeneration, i.e., revive the cryopreserved person.

While whole bodies can be cryopreserved, the main interest is in the brain, viewed as containing the essence of who we are in the stored memory and personality information. As stated by Alcor,

> If a brain can be preserved well enough to retain the memory and personality within it, then restoring health to the whole person is viewed as a long-term engineering problem.[6]

Whole body preservation costs more, at about 200,000 US dollars, and so some members choose "neuropreservation" (head only) for about 80,000 US dollars which, of course, is still out of reach for many people.[7] High fidelity preservation of the memory and personality information in the brain comes with the expectation that future biomedical technology will enable regrowing a body.

[2] Cryonics Institute—Technology for Life. https://www.cryonics.org/.

[3] http://www.oregoncryo.com/index.html.

[4] KiroRus. https://kriorus.ru/en.

[5] Alcor Life Extension Foundation, "The Alcor Team." https://www.alcor.org/AboutAlcor/meetalcorstaff.html.

[6] Alcor Life Extension Foundation, "Scientists' Cryonics FAQ." http://www.alcor.org/sciencefaq.htm. A good bit of technical information about cryonics is found in this section of the website.

[7] These are Alcor's charges. The Cryonics Institute has lower prices. See "The CI Advantage." https://www.cryonics.org/the-ci-advantage/.

Although polemical, one is hard-pressed to find a better succinct statement of some of the pros and cons of cryonics, from a consumer's perspective, than this one on the Alcor "Frequently Asked Questions" page. Responding to the question, "Why haven't more people signed up for cryonics?" the author of the page explains,

> People don't sign up for cryonics because it's not traditional, they're skeptical of anything they haven't seen work, it costs money, they're afraid of what their friends might think, they live in denial of their own death, they don't want to think about the subject, they procrastinate, they don't like life well enough to want more of it, or they are afraid of a future in which they may be alienated from friends and family and a familiar social environment. Typical Alcor members (if any Alcor member could be called "typical") tend to be highly educated independent minded people who enjoy life and think cryonics has a reasonable chance of working. They pay for it with life insurance and think the future is likely to work out pretty well. They often have friends or relatives who are Alcor members. They expect Alcor to revive them using nanomedicine and expect to continue their lives with as much passion and joy as today—only with much more amazing technology.[8]

A Brief History

Cryonics is not a new idea. American founder and inventor, Benjamin Franklin, speculated about preserving people in a suspended state.[9] Serious efforts to make it happen is traced to the 1962 book, *The Prospect of Immortality*, by Michigan College physics professor, Robert C. W. Ettinger.[10]

As the stories go, in the early days of cryonics, advocates placed their friends' bodies (and sometimes their pets) in the freezer locker, hoping for the best. The problem with literal freezing is that when ice thaws it expands. So, a frozen brain is destroyed upon thawing. The science of cryonics has progressed considerably. Now, a chemical process called vitrification (ice-free preservation) is utilized to attempt structural preservation of the brain for later "thawing." In a sense, vitrification is an extension of traditional and current funeral practices, which attempt to arrest decay with formaldehyde, hermetically sealed caskets, and cement vaults.[11]

Since 1967, nearly 200 people have been cryopreserved at Alcor, and about 1300 have made legal and financial arrangements to be preserved in the

[8] Alcor Life Extension Foundation, "Frequently Asked Questions." https://www.alcor.org/FAQs/faq04.html.

[9] *The Works of Benjamin Franklin*, vol. 6, ed. Jared Sparks (Chicago: Townsend Mac Coun, 1882), 382–83.

[10] Published by The Cryonics Institute.

[11] Oliver Krüger, "The Suspension of Death. The Cryonic Utopia in the Context of the U.S. Funeral Culture," *Marburg Journal of Religion* 15, no. 1 (2015): 1–19. See also Gary Laderman, *The Sacred Remains. American Attitudes towards Death, 1799–1883* (New Haven: Yale University, 1996).

future.[12] The numbers of cryopreserved people and those who have made legal and financial arrangements are difficult to determine for the rest of the industry, since many cryonics' organizations are small and do not publish their statistics, and the American Cryonics Society does not provide statistics.[13]

In recent years, Alcor and other cryonics organizations have worked hard to educate the public, appeal to the scientific community, and, as one might expect, confront various legal challenges.[14] The Alcor website offers scientific journal articles supporting cryonics and a "Scientists' Open Letter on Cryonics," with 68 signatories.[15] Here is the preface to the statement:

> *Signatories encompass all disciplines relevant to cryonics, including Biology, Cryobiology, Neuroscience, Physical Science, Nanotechnology and Computing, Ethics and Theology.* The signatories, speaking for themselves, include leading scientists from institutes such as MIT, Harvard, NASA and Cambridge University to name a few.

Here is the statement. It is followed by the names and brief credentials of the signatories.

> Cryonics is a legitimate science-based endeavor that seeks to preserve human beings, especially the human brain, by the best technology available. Future technologies for resuscitation can be envisioned that involve molecular repair by nanomedicine, highly advanced computation, detailed control of cell growth, and tissue regeneration. With a view toward these developments, there is a credible possibility that cryonics performed under the best conditions achievable today can preserve sufficient neurological information to permit eventual restoration of a person to full health. The rights of people who choose cryonics are important, and should be respected.

Our goal is not to defend cryonics nor suggest that it will be successful. It is not accepted by mainstream science. That said, it does have enough support to merit exploration, at the very least as a thought experiment.

When You Are Dead You Are Dead—Or Not

A major concern of Alcor and other cryonics organizations is to preserve the "dead" body, and especially the brain, with as much integrity as possible. If the

[12] Alcor Life Extension Foundation, "Alcor Membership Statistics." https://www.alcor.org/AboutAlcor/membershipstats.html.

[13] Max More provides an excellent discussion of the challenge of obtaining accurate statistics. See Alcor Life Extension Foundation, "CEO Statement on Membership Statistics." https://www.alcor.org/blog/ceo-statement-on-membership-statistics/.

[14] Richard Huxtable, "Cryonics in the Courtroom: Which Interests? Whose Interests?" *Medical Law Review* 26, no. 3 (2018): 476–499.

[15] Alcor Life Extension Foundation, "Scientists' Open Letter on Cryonics." https://www.biostasis.com/scientists-open-letter-on-cryonics/.

brain turns to mush, then there is nothing to revive later. Cryonics views death through an "information-theoretic criterion," holding that death occurs when the biological structures that encode memory and personality are destroyed, such that this critical information cannot be recovered.

Death is, after all, partly an interpretation. Death can be pronounced when the heart stops beating or, minutes later, with the flatlining of brain impulses, i.e., brain death. Theories about what constitutes death (e.g., cardiac or brain stoppage) have evolved with medicine and are sometimes understood through religion.[16] Cryonics protocols are organized around preserving the clinically and legally deceased person as quickly and as thoroughly as possible, to increase the odds that revival can occur with good results. Information-theoretic death occurs several hours following clinical death. Dying is a process, and in the ideal case cryonics intervention procedures begin immediately after the heart stops beating and prior to brain death.

If an Alcor member is on their death bed, stand-by teams from Suspended Animation, Inc., a company contracted by Alcor, work with the traditional medical team, as much as possible, to be allowed to move the body to a cool-down state immediately after legal death is pronounced. The body is then transported to Scottsdale, Arizona, where the formal vitrification chemical process is performed in the lab at Alcor's headquarters.

Importantly, there is apparently some minimal flexibility about when death is pronounced by a doctor. Alcor's strong preference is that the cryonics member be pronounced legally dead as early as possible, so that the preservation process can occur with the least possible disintegration of the brain. Some Alcor members, who are terminal, travel to Scottsdale, Arizona to live in an assisted living facility close to the company lab, so their vitrification process can begin as soon as possible after legal death. In fact, some members of cryonics' organizations want to control the death process totally, so they engage in cryothanasia, physician or self-administered euthanasia for the purpose of being cryopreserved. Alcor understands legally dead members to be "patients" and takes responsibility for their patients seriously. They keep the cryopreserved patients in a controlled environment and guard them around the clock. Within that controlled environment are the patients housed in dewars, each holding four full bodies and nine neuropatients.

Professor Mercer presented a paper at Alcor's national conference several years ago.[17] He had the opportunity to interact extensively with the members. Professor Mercer expected most members would be older and interested in cryonics in order to address their own coming demise. "Aging hippies" wanting to live forever were among the attendees. However, many members were much

[16] S. M. Setta and S. D. Shemie, "An Explanation and Analysis of How World Religions Formulate Their Ethical Decisions on Withdrawing Treatment and Determining Death," *Philosophy, Ethics, and Humanities in Medicine* 10, no. 6 (2015). https://doi.org/10.1186/s13010-015-0025-x.

[17] "Cryonics and Religion: Friends or Foes?" *Cryonics* 29, no. 1 (2008): 10–21.

younger, in their twenties and thirties. At first this was puzzling, but after conversing with some of these younger attendees, Professor Mercer theorized that since we have discovered all the "new" worlds and been to the moon, conquering death is the next exciting frontier. Indeed, many of the attendees, and those presenting at the conference, were from technical and professional fields—doctors, computer scientists, and information specialists—or were aspiring to those related fields of study.

Although the procedures are still being refined, and no cryopreserved person has yet been brought back, the plan is to revive the cryopreserved person when the cause of legal death can be remedied by medical science and the patient can be restored to full health. Skeptics say they will sign up when the first cryopreserved person is brought back safely. Alcor responds, "Would you rather be in the experimental group, or the control group?"[18]

Religious Issues

A cryopreserved person can be regarded as one would regard a coma patient. Just like the coma patient who may eventually revive, the cryopreserved person is understood to be in some sort of deep unconsciousness.[19] Cryonics, then, can be viewed as an extension of current medical procedures that treat coma patients. From this perspective, cryonics may raise no religious concerns, or at least no more than are raised by comas.

Alcor takes this approach to cryonics, stressing that cryonics is a technology not a religion, and the company has no philosophical or religious agenda. The company says cryonics is not resurrection and does not bring immortality, but that cryonics is consistent with the life affirming views of religion. Readers are referred to a number of positive statements made by religious writers.[20]

However, since cryonics preserves a patient for an indefinite period of time, and there are complicated social and legal issues involved, cryonics is likely not to be accepted by the religions as traditional medical procedures are, at least any time soon. The indefinite "purgatory" period would typically be a very long time, compared to a coma patient, because no one is close to figuring out how to bring a cryopreserved patient back. Some cryonics members are religious and interpret cryonics through the lends of their particular theology. However, the cryonics industry, and Alcor as the prime example, eschews a religious interpretation of their work. Our goal in this section is to reflect on how the religions might eventually interpret cryonics.

[18] Alcor Life Extension Foundation, "What is Cryonics." https://www.alcor.org/AboutCryonics/index.html.

[19] Alcor Life Extension Foundation, "Cryonics Myths." https://www.alcor.org/cryomyths.html.

[20] Alcor Life Extension Foundation, "Frequently Asked Questions." https://www.alcor.org/FAQs/faq04.html.

Monotheistic Religions

In this chapter, we consider one religion, Christianity, in some depth, as an example of how a religion could intersect with cryonics. We mention other religions but do not address them in the same depth. Our hope is that you the reader, and those interested in, or followers of, other religions, will use this thought experiment as a catalyst to make more connections to other faith traditions.

Resurrection

Resurrection is, understandably, not the way Alcor understands cryonics. From the perspective of religion, however, resurrection is an interesting theological prism through which to view this procedure. While Judaism, Christianity, and Islam have doctrines of resurrection, Christianity brings this idea to center stage. We will think about cryonics in light of resurrection, with particular, though not exclusive, reference to Christianity.[21] Karmic religions have stories and concepts of resurrection, but the idea is not as central as it is to the Christian religion.

The Greek for resurrection is *anatasis*, literally "stand" (*stasis*) "up" (*ana*). In the biblical tradition, resurrection can take two forms. In resuscitation of the dead, to play on the Greek word, we "stand up" as embodied in the same kind of body we had before, except that the disease that killed us is miraculously gone. We continue living a normal life until the next disease or accident comes along. As an example, Christians believe that Jesus raised Lazarus from the dead,[22] and Lazarus presumably continued with his life pretty much as before. This resuscitation of a dead body is a different kind of raising than resurrection to eternal life with God, with new abilities and possibilities. We have coined the term "transformational resurrection" to refer to this kind of raising, which could also be called eschatological (i.e., end-time) resurrection.

Cryonics lends itself to being understood as technological versions of either type of resurrection. In the first version, resuscitation of the dead, the cryopreserved body is restored to full functioning by using the now available medical treatment to cure whatever killed the person before cryopreservation. We are using the terms "dead" and "killed," but keep in mind that the cryonics industry does not believe death has occurred. The issue for Alcor, for example, is reviving a person who is a patient waiting in their cryopreserved state until resuscitation. Following revival, the person continues life in a body, now repaired but in effect a continuation of their pre-cryopreserved body.

[21] Much of the material in this section is adapted from Calvin Mercer's articles, "Resurrection of the Body and Cryonics," *Religions* 8/5, 96 (May 2017): 1–9. www.mdpi.com/journal/religions. https://doi.org/10.3390/rel8050096, in a special issue "Religion and the New Technologies." http://www.mdpi.com/journal/religions/special_issues/new_technologies; and "Cryonics and Religion: Friends or Foes?"

[22] John 11.

The more dramatic and interesting version of resurrection is a transformative event that brings one into new and eternal life with God. If the revival of the cryopreserved patient occurs far enough into the future, the technology may be developed to transform the patient into a new being, enhanced in abilities far beyond the person who was pronounced legally dead and then cryopreserved. Transformative resurrection in the biblical tradition entails at least three important aspects: embodiment, transformation, and continuity of personal identity. To be compatible with the religion Christianity, cryonics probably needs to satisfy at least these three criteria.

Embodiment

The Apostle Paul was arguably the most important first-century author and leader responsible for the foundational ideas of the Christian religion. When interpreting Paul, or for that matter any early Christian writer, it is important not to over-read. Paul was not a systematic theologian, sitting comfortably in his faculty office and having convenient access to a huge research library. Paul was on the front lines as a missionary and church organizer. On the road a lot and sometimes thrown into jail, his letters to churches were written for specific purposes, usually to address particular troubling issues in faith communities, usually but not always ones he had organized.

In the Christian Testament, also called the New Testament, Paul's clearest presentation of resurrection came in his first letter to the Corinthian church and especially chapter 15. Paul was influenced by both his Hellenistic and Jewish cultural and religious backgrounds. With regard to what we can call theological anthropology (or what it means to be human, from a theological perspective), however, Paul's view is grounded mainly in his Jewish background. The normative Jewish view is that a person is a psychosomatic unity. We are not various parts—body, soul, spirit—somehow stapled together and that can be dissected. Rather, in a Jewish view, we are a whole, and so resurrection is not of a soul, but, rather, of the whole person, including the person's physicality or embodiment. We provided background to this view in the chapter, "Transhumanism, the Posthuman, and the Religions."

Regarding his theological anthropology, the most comprehensive Greek term used by Paul to describe being human is *soma* (body). For Paul, *soma* is intimately related to the person as a whole and does not mean our physical dimension only. So, while *soma* emphasizes our physicality, this physicality does not and cannot stand alone in any sense. In a significant theological move, Paul combined *soma* with *pneumatikon* (spiritual) in, for example, 1 Corinthians 15:44, to name the resurrected existence "spiritual body" (*soma pneumatikon*).

In addition to being influenced by his Jewish background, Paul models his notion of bodily resurrection on Christ,[23] who is described as the "first fruits"

[23] e.g., Romans 8:11; 1 Corinthians 6:14.

of those who have died.[24] Christians understand that Christ was raised as a whole person, not merely as a disembodied soul. That the tomb was physically empty on Easter morning has important theological implications. It is believed that the bodily resurrected Christ had supernatural abilities, appearing through closed doors[25] and suddenly on a remote road[26] in the post-resurrection stories.

Pauline thinking is not the only way to conceptualize Christian anthropology. The Hellenistic Greek tradition, which differed considerably from the Jewish, influenced some strands of early Christian theology. Some Christians today, especially those with conservative leanings, think about the constitution of the person and the believer's resurrection in terms more akin to immortality of and transmigration of the soul, notions prominent in some karmic religious traditions. In this view, detailed in the chapter, "Transhumanism, the Posthuman, and the Religions," the soul is a part of the person that survives after death of the body. Indeed, "drop the body" is a phrase sometimes used in Asian culture for death.

It is interesting to think about whether the intermediate state between death of the individual and the "last day" (when resurrection is believed to occur) can be related to the time period between cryopreservation and later restoration. The Bible and the ancient extra-biblical books in the *Apocrypha*[27] speak some about this interim period. During this interim state, purgatory is a time to accomplish the goal of purification and preparation for the heavenly age to come, loosely analogous to the transhumanist goal of enhancing us for life long into the future. We should note that not all Christians believe in purgatory, and some see purgatory as punishment.

Finally, we noted earlier that cryonics offers two options for cryopreservation, whole body preservation or neuropreservation (head only). The contention that we have "body memory" and body knowing, apart from information solely in the brain, would argue against head-only preservation.[28] Artists, musicians, and athletes often report a sense that their hands or other body parts are functioning in addition to or even separate from their conscious thought processes. Whether the mind extends beyond the brain is a complicated philosophical and theological question; we will touch on this topic in the chapter on mind uploading.

Transformation

Transformative resurrection is a type of raising that goes far beyond mere resuscitation of a dead body. Transformative resurrection means we die and are

[24] 1 Corinthians 15:20–23.

[25] John 20:19

[26] Mark 16:12; Luke 24:13–32.

[27] e.g., 2 Maccabees 12:39–45.

[28] For a helpful discussion about body memory and theology, see John Swinton, "What the Body Remembers: Theological Reflections on Dementia," *Journal of Religion, Spirituality & Aging* 26 (2014): 160–172.

raised to new, redesigned, revitalized, and enhanced life. As we have seen, Paul's term for this resurrected life is *soma pneumatikon*, spiritual body.

The spiritual body Paul envisions, however, does not continue in the same way the pre-resurrection life did as before death. Again, the model is Jesus,[29] who was raised to a transformed existence, qualitatively different from the pre-resurrection body. New possibilities abound. Jesus appeared and disappeared before witnesses,[30] moved through doors,[31] and became invisible.[32] The believer's raised body, modeled after the resurrected Christ, is believed to be imperishable, glorious, and powerful. Typical "flesh and blood" it is not. The perishable, dishonoring, and weak will be transformed.[33] Paul says, "He will transform the body of our humiliation (could be translated, "humble bodies") that it may be conformed to the body of his glory (could be translated, "his glorious body).[34] In early Christian end-time thinking, the believer's transformation is also part of the anticipated cosmic transformation of a "new heaven and a new earth."[35]

Curious minds can certainly come up with many questions at this point. What happens when someone dies by being burned up completely in a fire, or what about cremation? Are the various atoms of that body, now floating all over the earth, somehow brought back together in the resurrected spiritual body? Is not the concept of a spiritual body an oxymoron? If spiritual means non-physical, then in what sense is there a body?

The Corinthians, to whom Paul was writing, were asking the same kinds of questions. "With what kind of body do they come?"[36] Paul's response to the curious Corinthians in the very next verse was: "Fool!"[37] Paul is not an engineer or physicist working out all the interesting particularities of his theology of resurrection. Paul is making theological statements, in this case affirming the psychosomatic unity of the person, in accord with his Jewish background.

So, one interpretation of the believer's resurrection is that Christians (and perhaps others outside the religion, depending on the salvation theory embraced) will be raised after death in a spiritual body, radically transformed with enhanced capabilities. A feasible religious interpretation is that cryonics provides the technological means for how God accomplishes transformational resurrection.

When the cryopreserved person is restored, probably far into the future, we may then have a menu of tools, such as robotics, tissue regeneration, and nanotechnology with which to cure and enhance the revived person. Whatever

[29] 1 Corinthians 15, Romans 8:11.
[30] Luke 24:13–43.
[31] John 20:26.
[32] Acts 9:1–9.
[33] 1 Corinthians 15:42–43.
[34] Philippians 3:21.
[35] Revelation 21:1.
[36] 1 Corinthians 15:35.
[37] 1 Corinthians 15:36.

that future restored body might look like, we can safely say that, compared to our bodies now, the future version will deserve to be called "transformed." In the chapter, "Transhumanism, the Posthuman, and the Religions," we saw that both the Mormon Transhumanist Association and the Christian Transhumanist Association support and call for modern enhancement technology, as a crucial part of God's plan.

By no means are we suggesting that the ancient scriptural depictions of resurrection are literally and directly anticipating technological developments in the scientific age. They are not. We are suggesting, however, that Paul's presentation of transformational resurrection can inspire the monotheistic religions to reach for a vision that remains true to key scriptural and theological principles while incorporating rapidly advancing enhancement technologies.

Some theologians might object that a newly grown body or a robotically enhanced body would not be made of the same kind of stuff as that constituting the person's body prior being cryopreserved. Paul, however, clearly saw the resurrected spiritual body as both consistent with the Jewish insistence on physicality and as different in that the resurrected spiritual body had capabilities not afforded to the pre-death body. Our point is that if Paul's spiritual body, constituted differently than the pre-death body, was theologically acceptable, then perhaps many strands of the Abrahamic religions might find the revived cryopreserved body, enhanced via robotics or nanotechnology, acceptable.

Personal Identity

Although radically transformed, Jesus' identity was not lost. Those who knew him before the resurrection knew and recognized him after. The disciples thought they were seeing a spirit, but Jesus said, "See my hands and my feet, that it is I myself; handle me, and see; for a spirit has not flesh and bones as you see that I have."[38] Jesus told Thomas, "Put your finger here, and see my hands; and put out your hand, and place it in my side."[39] The pre- and post-death being is like a kernel sown into the field and which grows into a body consistent with the kind of seed sown.[40]

This feature of Paul's vision of the resurrection is the least problematic in light of a restored cryopreserved person. The philosophical and legal issues may be more complicated, and we discuss them in the upcoming mind uploading chapter, but Alcor's goals and contracts envision that the restored person is a continuation of the individual who legally died and was cryopreserved.

Scriptural and Theological Objections

Even if cryonics figures out how to successfully restore a cryopreserved person in a way that meets the three criteria of physicality, transformation, and

[38] Luke 24:39.

[39] John 2:27.

[40] 1 Corinthians 15:35–41.

continuation of personal identity, other scriptural and theological objections can be made.

In the resurrection, Christianity promises the new creation of God will live in the blissful arms of God, free from sin and suffering. The transhumanist future, which may include the restoration of cryopreserved people, is anything but guaranteed to bring in an age of bliss. In fact, as we have seen in other chapters, serious concerns are being expressed that radical enhancement technologies may bring all kinds of mischief. Cryopreserving and restoring a Hitler simply perpetuates evil.

We know of no indisputable response to this concern. We can reasonably expect that advances will be made in many enhancement domains as we move into the future. One can hope that moral and spiritual enhancements will bring us closer to a more virtuous, if not sinless, future by the time we are able to restore cryopreserved individuals. Maybe the cryonic restoration process would include moral and spiritual enhancements, along with the physical, cognitive, and emotional repairs and enhancements. Basing the future prospects of the world on this hope may very well sound scary and unlikely, so the objection is a legitimate one.

People of faith might also object to cryonics by asking why the faithful should resist death so strenuously if life after death with God awaits. The response depends, at least in part, in how this is broken out into liberal versus conservative responses. Conservatives focus more on otherworldly concerns; hence, this objection may carry weight with them. Liberals, however, focus on this-worldly concerns and might be more inclined to see life as a divine gift that we are to protect as we work to conform to God's will and seek justice here on earth.

It complicates the analysis, but likely muddled into the equation is a fear of death that many of us have absorbed in our death-denying and death-defying culture. A middle way forward on this point for people of faith is that we are neither meant to rush toward death nor to avoid death at any cost. Cryonics could be seen by religion as one potential way to preserve life as long as cryonics is not the only option. Death, too, is an option and, from a religious point of view, death is not the end.

The attempt to show parallels between resurrection and cryonics also runs into the objection that transformative resurrection results in eternal life and cryonics' technological restoration is to a physical body that is still subject to decay or trauma. This objection is easy to counter in part. The kind of radical technological restoration that might be coming is going to be ongoing, in the sense that cryopreservation would always be an option should it be needed again to give the person time to await whatever additional restorative capabilities are needed.[41] Of course, an accident or other trauma could damage the

[41] The discussion of LEV, the Longevity Escape Velocity, in the chapter, "Superlongevity," is relevant here.

embodied person beyond any possible reconstruction such that cryopreservation is not possible.

Common ground between religion and cryonics that allows for wholesale embrace of cryonics by the faithful may never be found. However, both the religious faithful and cryonics advocates agree, at least in general terms, that death is not necessarily the final word and that there is hope beyond. Perhaps both groups could agree with this vision asserted by Paul in the Christian scriptures:

> When this perishable body puts on imperishability, and this mortal body puts on immortality, then the saying that is written will be fulfilled: "Death has been swallowed up in victory. Where, O death, is your victory? Where, O death, is your sting?"[42]

Cryonics advocates look to technology and future medicine to fulfill that hope beyond legal death. The religious faithful depend upon God for that hope. Whether or not cryonics will be seen as a way religious hope can be realized is a debate that will continue.

Karmic Religions

Reflection on the intersection of cryonics with the monotheistic religions is scarce enough. Unfortunately, we have even less analysis regarding the intersection of cryonics with karmic religions, so we can only begin to speculate on how that dialogue might go.

A possible objection to cryonics concerns one's motivations for seeking preservation. Karmic religions are concerned about attachment to the body as constituent of one's identity. Using Buddhism as an example, this religion views all phenomena as impermanent. The desire to identify with the impermanent body yields suffering. So, to the degree that interest in cryonics is fueled by a desire to maintain the body, it is an unhealthy desire that hinders spiritual progress.

Finally, the karmic notion, as in Hinduism, that we have a soul, can fit loosely into a cryonics scenario. Cryonics is concerned with the preservation of who we are, identified as our memory and personality. If soul can be understood as a word that entails who we are in our deepest selves, then cryonics is about preserving that into future bodies. Neuropreservation (i.e., head only) is particularly compatible with the idea that there is some "part" of us, located in the brain, that is our essence and can be transferred to future "incarnations."

[42] 1 Corinthians 15:54–55.

Ethical Issues

What It Means to Make Us Better: The Therapy—Enhancement Continuum

In this framing of the issue, therapeutic interventions are generally accepted if the risks do not outweigh the possible benefits. Enhancements, however, raise concern, and the more extreme the enhancement the more concern. Key to the analysis in this section is the location of cryonics on the continuum and religious views of what makes enhancing technology good and, therefore, potentially acceptable. Recall that "normal" is not necessarily the dividing point between acceptable and unacceptable on this continuum. For Judaism and Christianity, the dividing point may be the point at which we are perceived as no longer living as a created image of God.

Cryonics' advocates often attempt to position cryonics on the therapeutic, and therefore acceptable, side of the continuum. They might suggest, for example, that cryopreserved individuals are patients and that being cryopreserved is somewhat like being in a coma. A significant difference is coma patients do not choose to be in a coma, although medically induced comas may be an exception to that. Still, any contention that cryonics is an extension of current life-saving medicine helps normalize cryonics.

Critics contend that using chemicals to supposedly save a legally dead person for perhaps centuries and then try to revive them is, on its face, extreme and anything but therapy. Despite its pro-healing stance, religion does not automatically support any medical intervention, especially if the risks are sometimes judged to outweigh the possible benefits, and extreme interventions that may not yield healing results are highly controversial. There are limits on how far to go to preserve life. For example, there is debate in the religions about "do not resuscitate" orders (DNRs), nutritional gastro-intestinal tubes when the patient cannot eat or drink, respirators to sustain breathing, and other artificial means of keeping people alive. Intense debate also surrounds using measures to hasten death.

The karmic religions are interested in generating the most good karma, thereby enhancing one's spiritual state in the next life. Cryonics is a moot point for critics of cryonics who believe the "patients" are actually dead. In this case, the soul has already moved on. For followers of the karmic traditions who do accept the premise of cryonics, the issue becomes how much and what kind of karma is generated from choosing cryonics.

Followers of the monotheistic faiths ask what action is in keeping with divine will; the answer usually does not involve extreme measures. Roman Catholic Christians, for example, are usually opposed to extreme measures to sustain life, although once extreme measures have been started, then generally one is not allowed to hasten death by stopping these measures.[43]

[43] A. Mackler, *Introduction to Jewish and Catholic Bioethics* (Washington, D.C.: Georgetown University Press, 2003).

The definition of death and regard for the body will be important factors in assessing the ethical acceptability of cryonics. Because of the sanctity of the human body, it is difficult to imagine that most Jewish people would permit neuropreservation. Cryopreservation as well presents the complication of not being certain about death, since revival is not assured. Some rabbis see death as occurring only if a return to life is impossible. Others interpret death as the irreversible cessation of respiration. Cryonics puts the question of irreversibleness up in the air.

Certain rituals also come into play and obviously depend on whether one is believed dying or is dead. In Judaism, sometimes the dying person shares what is called an ethical will in which they identify values and guidance from their journey in life. The confession of sins just before death is a traditional practice in Judaism and Christianity. Ritual purification and accompaniment of the body immediately following death is important in Judaism. All of these practices are complicated by cryonics. For example, should one confess their sins in case later attempts at revival are unsuccessful?

In the Abrahamic religions, divine sovereignty, however it is specifically interpreted, usually means that people belong to God. Our lives are on loan from God, and humans are charged with being good stewards of their finite lives. Cryonics can be seen as "playing God," as crossing the line from what it means to be created in God's image as a mortal, limited creature, to a risky, unproven bid to enhance beyond the divine mandate. The comeback from religious followers who choose cryonics is that they are honoring God's creation by using God-given technology to preserve and safeguard the divine gift of life.

Cryonics is new ground. Religions will have to decide if cryonics is sustaining life or only complicating death, especially if more people become interested in cryopreservation.

Choice

Choosing cryonics may simply be seen as a far-fetched way to spend lots of money or a desperate attempt to avoid dying. Either way, we may shrug our shoulders and say it is up to the individual to choose. But, as pointed out many times in this textbook, individual choices affect far more than the individual making the choice.

Potential ethical implications of a choice for cryopreservation include the allocation of a significant amount of money to be cryopreserved and possible relational issues. We have discussed the financial issue in other contexts. At this point, cryonics is not chosen by large numbers of people. For those who do choose it, there are likely to be family members with a variety of concerns about finances, false hopes, disposition of the family member's body, and others. Depending upon the significance of the relevant rituals practices and beliefs of those affected, these concerns in themselves may or may not be sufficient to make a choice for cryonics unethical. The point is that choices we make, including about cryonics, impacts others, often significantly. Good ethical reasoning takes this impact into account.

If cryonics were to be made widely available, new questions will arise about choice. Perhaps everyone should have an equal right to be cryopreserved. We can anticipate, however, that a counterargument would be made. Should mass murderers, for example, be allowed use of the technology, and who should get to decide such questions?

Justice

We have seen that a common criticism of radical enhancement technologies is that, even if acceptable on other grounds, they will be the privilege of the wealthy and powerful class. With the expense involved in both whole body cryopreservation (200,000 US dollars) and neuropreservation (80,000 US dollars),[44] a concern about the just distribution of opportunities has been raised. The religious traditions, especially in their more liberal versions, insist that life-giving opportunities be fairly distributed.

Two responses can be made to this criticism. First, advocates of cryonics contend that individuals and societies constantly make decisions about expensive medical treatments, and cryonics should not be viewed any differently. In other words, if an individual with the financial means chooses to expend their resources on cryonics, they should have that right, just as another individual should have the right to expend their resources on an expensive heart transplant procedure. This response, comparing cryopreservation to other medical procedures in cost, does not address the distributive justice concern. It just underscores the widespread nature of inequities and related social and distributive justice issues.

A second response about the cost of cryonics is that the industry has actually developed a way that some low-income people can afford to enroll to be cryopreserved. Life insurance is relatively inexpensive, and affordable, for most young people who are healthy. So, access to the expensive whole body preservation or neuropreservation procedures can be achieved by paying for life insurance and signing over the policy to a cryonics company, such as Alcor or Cryonics Institute. Alcor claims that most of its members are middle-class and are funding cryonics through life insurance.[45]

This creative approach to funding expensive cryopreservation does address the distributive justice concern, but only for one population, namely, young, healthy individuals. The concern about prohibitive costs remains for all other low-income, low-wealth people, and the result is a familiar one for expensive radical enhancements—the wealthy class lives longer and becomes even more powerful.

[44] These are Alcor's costs, as discussed earlier in this chapter.

[45] Alcor Life Extension Foundation, "Cryonics Myths." https://www.alcor.org/cryomyths.html#myth4.

With cryonics, since no cryopreserved person has been revived, the future of that person upon revival is unknown, making the benefit-risk calculation harder. There are a number of possibilities. The ideal scenario is that the person is revived successfully, with their mental faculties intact, and the technologies that have become available since that person was cryopreserved can provide for them a healthy life of indefinite length. Perhaps the renewed life can come with options of cognitive, affective, moral, and spiritual enhancements. That possibility is justification enough for many who embrace a proactionary stance. It may also be the case that cryopreservation quickly, or at least somewhat quickly, becomes available and accessible to people of diverse social classes and geographical location.

Concerns from a precautionary perspective include the risk of the revival falling short, perhaps far short, of the ideal scenario. Perhaps the revival is compromised in ways that leave the person with significant cognitive, affective, or other deficits that entail great suffering. Or, perhaps the person is revived fully functioning, but finds themselves in a world to which they cannot successfully adapt, because of financial or other challenges. Keep in mind that it may be hundreds or thousands of years before cryopreserved people can be successfully revived, if ever, and the world will have changed considerably in that time. And, it may be that significant aspects of the person's identity are lost if identity turns out to extend beyond the brain. There is also the risk of increased social and distributive injustice.

Questions for Discussion

1. Would you consider cryonics for yourself? Why or why not?
2. If you would not consider cryonics, would your answer change if your close family members chose cryonics?
3. When do you think death occurs? Why do you think this?
4. How would you differentiate spending tens of thousands of dollars on being preserved through cryonics from spending that money on a second home in the mountains, cosmetic surgery, food for those who are in need, or a needed heart transplant? Consider your top five values in your response.
5. Why do you think younger generations are interested in cryonics?
6. What challenges, other than the ones we have discussed, can you identify for someone who is cryopreserved today and revived, for example, 1000 years from now?
7. Can you think of any other relevant religious arguments either for or against cryonics not discussed in this chapter?
8. Reflect on how you think cryonics, as a thought experiment, might help us to re-consider religious thinking. Identify examples.

CHAPTER 9

Mind Uploading: Cyber Beings and Digital Immortality

Technology

Whole Brain Emulation

"Whole brain emulation" is the technical term for what is commonly called "mind uploading" and sometimes "mind copying" or "mind transfer." It refers to copying the information (i.e., memory and personality) in the brain into a digital substrate (part of a computer). Mind uploading is the stuff of science fiction that has made its way into serious conversation about transhumanist possibilities.

Calling this procedure "mind" uploading treads into very complex questions about the definition of mind. Mind can be understood as a general term, a concept, that refers to particular mental states, such as desire, emotion, perception, intention, belief, and others.[1] We will address, in a limited way, philosophical questions as they are useful in our theological conversations. This chapter is not a thorough introduction to all the philosophical issues relevant to whole brain emulation. Our purpose is to stimulate thinking about the intersection of mind uploading and religion, drawing on some philosophy to help shed light on this intersection.

Although major technical barriers must be crossed, and there is plenty of disagreement about the scientific feasibility of mind uploading, some thoughtful critics assert mind uploading will be achievable at some point. Although spoken over three decades ago, Hans Peter Moravec, associated with the Carnegie Mellon University Robotics Institute, articulated a view still held by many transhumanists, that a human being is

[1] See Gilbert Ryle, *The Concept of Mind* (Chicago: University of Chicago Press, 1949); and David Armstrong, *A Materialist Theory of the Mind* (London: Routledge & Kegan Paul, 1968).

C. Mercer, T. J. Trothen, *Religion and the Technological Future*,
https://doi.org/10.1007/978-3-030-62359-3_9

> ... the *pattern* and the *process* going on in my head and body, not the machinery supporting that process. If the process is preserved, I am preserved. The rest is mere jelly.[2] (italics are original)

Ray Kurzweil's is a familiar transhumanist vision along the same lines:

> Up until now, our mortality was tied to the longevity of our *hardware*. When the hardware crashed, that was it... As we cross the divide to instantiate ourselves into our computational technology, our identity will be based on our evolving mind file. *We will be software, not hardware...* When the hardware is trillions of times more capable, there is no reason for our minds to stay so small. They can and will grow. As software, our mortality will no longer be dependent on the survival of the computing circuitry. There will still be hardware and bodies, but the essence of our identity will switch to the permanence of our software.[3] (italics are original)

How to Transfer a Mind

In a previous chapter, we introduced Nick Bostrom, a philosopher who directs the Future of Humanity Institute at Oxford University. In addressing the technical aspects of mind uploading Bostrom details the several steps required for mind uploading.[4] In summary, a thorough scan allows for reconstructing the three-dimensional neuronal network of a brain. This network information is combined with models of how the neurons function and then is placed in a powerful computer, providing an active digital simulation of the scanned brain.

Basically, we scan the structure of a brain and construct a software model of it that would behave essentially the same way as the scanned brain. Experimental work is already occurring on smaller animal brains. Incremental technical advances mean lower grade emulations will prepare the way for higher grades and, perhaps eventually, high-fidelity emulations. Bostrom admits that a good bit of technical progress, although attainable, is yet needed for success with humans. But, he says,

> The whole brain emulation path does not require that we figure out how human cognition works or how to program an artificial intelligence. It requires only that we understand the low-level functional characteristics of the basic computational

[2] Hans Moravec, *Mind Children: The Future of Robot and Human Intelligence* (Cambridge, MA: Harvard University, 1988), 117.

[3] Ray Kurzweil, *The Age of Spiritual Machines: When Computers Exceed Human Intelligence* (New York: Penguin, 1999), 128–29. See also Ray Kurzweil, *How to Create a Mind* (New York: Viking, 2012).

[4] See his book, *Superintelligence: Paths, Dangers, Strategies* (Oxford: Oxford University Press, 2014). See also Randal A. Loene, "Feasible Mind Uploading," 90–102, in Russell Blackford and Damien Broderick, eds. *Intelligence Unbound: The Future of Uploaded and Machine Minds* (West Sussex, UK: Wiley Blackwell, 2014); and Robin Hanson, *The Age of Em: Work, Love and Life When Robots Rule the Earth* (Oxford: Oxford University, 2016).

> elements of the brain. No fundamental conceptual or theoretical breakthrough is needed for whole brain emulation to succeed.[5]

With the tagline "Promoting R&D for Whole Brain Emulation," Dutch neuroscientist and neuroengineer Randal A. Koene founded the company, Carboncopies.[6] On the website, it is admitted there is much to accomplish before mind uploading can be achieved, but calls whole brain emulation "the most promising technological path to overcoming our fundamental limitations as a species." A good bit of technical information is provided on the website, in a way accessible to non-specialists.

Current medical technology places neural implants into the human brain. We have used these implants mainly for brain stimulation in the treatment of clinical depression or Parkinson's Disease. But other implants are being developed. Whole brain emulation adds a new direction to such efforts, potentially moving the "mind" into a computer—largely uncharted territory.

While moving (i.e., uploading) one mind into one computer is the usual vision of mind uploading enthusiasts, various uploading possibilities may unfold. The source (i.e., original) mind may survive, or it may be destroyed in the uploading procedure. Perhaps more than one upload is achieved. Maybe the mind is uploaded into the Cloud, providing for a different set of issues. Ray Kurzweil thinks that the mind uploading will happen so gradually that we will not notice the transfer.[7] The more cynical view that it is ridiculous to think this project will ever work, is shared by many and captured in in this blunt quote: Mind uploading is "nothing more than a novel way to commit suicide."[8]

Matthew Zaro Fisher articulates the varied possibilities at play between the extremes of a perfect upload and certain death. Fisher asks whether whole brain emulation:

> … would produce the person in his or her full sense of self-presence, produce an 'echo' of one's self-presence, produce a second person with a distinct self-presence (something like an echo-twin), or would merely be a pattern of data running with "nobody home."[9]

[5] Ibid., 30.

[6] Carboncopies. https://carboncopies.org/.

[7] Ray Kurzweil, *The Singularity is Near: When Humans Transcend Biology* (London: Penguin Books, 2005), 202.

[8] Nicholas Agar, "Kurzweil and Uploading: Just Say No!" *Journal of Evolution and Technology* 22 (2011): 27.

[9] "More Human than Human? Toward a 'Transhumanist' Christian Theological Anthropology," in *Religion and Transhumanism: The Unknown Future of Human Enhancement*, eds. Calvin Mercer and Tracy J. Trothen (Santa Barbara, CA: Praeger, 2015), 26. If mind uploading results in a new intelligent, sentient being, the religions will have to address whether and how to embrace them. For a discussion of this, see Calvin Mercer, "A Theological Embrace of Transhuman and Posthuman Beings," *Perspectives on Science and Christian Faith* 72/2 (June 2020) 1–6.

The basic idea in Fisher's quote is fleshed out in a blog post from Carboncopies, the research company introduced earlier:

> *The road to emulation is a gradual incline with many stages and milestones.*
>
> Undeniably, the first case of a successful emulated human mind will have a huge impact on the world and make its mark in our history books. It will also likely be far less sudden and unexpected than we might tend to imagine. Building an emulation is best done in stages. For example, treating each section of the brain as separate emulation projects would be easier to focus on than tackling the entire brain at once. There will be successful emulations of sections of the brain before they can operate as one complete operating substrate. ... Progress will be marked by emulations of animal brains with increasing cognitive complexity. This is why subjects such as the fruit fly drosophila and the nematode worm Caenorhabditis elegans are prime initial emulation subjects. It will also be a massive cooperation effort. Teams from all over the world are already working on various research and development projects that ultimately help get us to whole brain emulation. ... Therefore, when the day comes to witness the first successful human emulation it will be more like a sigh of relief for all the hard work rather than some grander surprise.[10]

Sometimes whole brain emulation is compared to "mindcloning," that is, creating a digital version of one's mind. "Mindware" software activates a "mindfile," a digital repository of memories, feelings, and thoughts.[11] A mindclone, while beginning as a replica of a human mind, may develop its own identity in the context of ongoing interactions and is, therefore, gaining attention in conversations about transhumanism.[12] While mindcloning may provide a historical record of who you are and perhaps software that can activate the record, if it stops short of an evolving, continuing consciousness it is not compelling to those who want to survive as a conscious personal identity.

> It seems that I gain no advantage unless I, Russell, obtain an extended period of life and other benefits. Perhaps it might be nice to think my digital twin will enjoy the benefits, but that is merely a consolation prize. The real prize is that I will be able to live much longer, perhaps even enjoying immortality, and that I will be enhanced in various other ways. If I'm to obtain the advantages of uploading, there must be some sense in which the upload is the original me, or in which I have survived as the upload.[13]

[10] Carbon Copies, "Recent Posts." https://carboncopies.org/what-will-the-first-substrate-independent-mind-look-like/.

[11] Martine Rothblatt, *Virtually Human: The Promise—and Peril—of Digital Immortality* (New York: St. Martin's, 2014).

[12] Steve Fuller discusses the work of Rothblatt, with regard to mindclones, in *Nietzschean Meditations: Untimely Thoughts at the Dawn of the Transhuman Era*,. Posthuman Studies 1, ed. Stefan Lorenz Sorgner (Basel: Schwabe Verlag, 2019), 118–24.

[13] Russell Blackford, "Introduction II: Bring on the Machines," in *Intelligence Unbound: The Future of Uploaded and Machine Minds*, 11–25, eds. Russell Blackford and Damien Broderick (West Sussex, UK: Wiley Blackwell, 2014), 17.

Religious Issues

In this section we address a number of complicated issues related to whole brain emulation. These issues are about what it means to be a human person, as understood by religious traditions. Much of the background for our discussion here was laid in the earlier chapter, "Transhumanism, the Posthuman, and the Religions." Mind uploading raises many of the same religious and ethical issues as cryonics. So, while we attempt to assess things with a broad religious lens, Christianity serves as the primary illustration of religious issues, and our discussion hopefully will serve as stimulus for an expansion of the conversation into other religions.

In the biblical tradition, a fundamental idea is that human beings are created in the image of God (Latin, *imago Dei*), first found in the ancient Israelite creation story[14] and a notion commanding much attention from biblical scholars and theologians through the centuries. The image of God has been understood in several ways,[15] giving emphasis to our rationality, free will, agency in the world, or some other feature. An interpretation that has come into favor is that the image of God means that human beings are relational, exhibited in our relationship with the creator God and with each other. Of course, the image of God can be multifaceted, incorporating more than one of these components. We will keep the image of God as an overarching theme as we work though this section.

As expressed by advocates, the main interest in whole brain emulation is to preserve who we are, our personal identity, into a host more reliable than our physical bodies. For Hinduism, to take one example of a karmic religion, the soul is the deep essence of who we are; the religious question has to do with whether mind uploading transfers the soul to the new platform. As complicated a theological question as that might be, Buddhist theology on this issue is likely to be even more labyrinthine, with its somewhat mystifying notion of "no-self."

Physicality

Mind uploading raises a host of complicated theological questions. The role of the body is a fundamental one, since transhumanists are often critiqued as viewing the person as patterns of information that are not necessarily tied to the biological body.[16] Relatedly, Jewish and Christian theologians, who affirm the importance of embodiment, are concerned about what they perceive

[14] Genesis 1:26–28.

[15] For a review of options presented in the context of transhumanism, see Jeanine Thweatt-Bates, *Cyborg Selves: A Theological Anthropology of the Posthuman* (Burlington: VT: Ashgate Publishing, 2012), 109–17.

[16] Secular advocates of whole brain uploading do tend to minimize the role of our bodies. See, e.g., Ray Kurzweil, *How to Create a Mind*, and Hans Morevac, *Mind Children: The Future of Robot and Human Intelligence* (Cambridge, MA: Harvard University Press, 1988).

(sometimes rightly) to be transhumanism's denigration of the body biological, therefore making some transhumanist projects like mind uploading theologically problematic.[17] In two previous chapters, "Transhumanism, the Posthuman, and the Religions" and "Cryonics," we gave some background regarding monotheistic religions and the importance of the created material world in general and the human body in particular in understanding personhood. As with cryonics, most reflection on mind uploading and religion has addressed monotheistic religions only.[18]

In the main, monotheistic religions, stemming from the biblical tradition, understand the body to be an essential aspect of a person. In other words, a human being is a psychosomatic unity of body and soul. We are not body, soul, spirit, mind, or other distinct parts somehow glued together. Rather, these words express dimensions or aspects of the holistic integrated person.

Indigenous traditions, perhaps more than any other way of being religious or spiritual, emphasize the interconnectedness of a person's multiple aspects, including spirit, affect, cognition, and physical being. The medicine wheel, an important part of Indigenous wisdom, illustrates how healing is understood as necessarily having multiple dimensions.[19] Healing from addiction, for example, is entwined with spiritual and cognitive healing. Many Indigenous people see traditional medicine as complementary to or blending with allopathic medicine.

With this theological anthropology as background, the concept of resurrection, found in ancient Judaism and most extensively expressed in early Christianity, is a bodily resurrection event. The whole person is raised, not just the soul. In previous chapters, we distinguished (1) resurrection as resuscitation from the dead and (2) resurrection of the believer to a transformed life

[17] See., e.g., discussions in Amy Michelle DeBaets, "Rapture of the Geeks: Singularitarianism, Feminism, and the Yearning for Transcendence," in *Religion and Transhumanism: The Unknown Future of Human Enhancement*, 181–98, eds. Calvin Mercer and Tracy J. Trothen (Santa Barbara, CA: Praeger, 2015), 183–84; Jeanine Thweatt-Bates, *Cyborg Selves: A Theological Anthropology of the Posthuman* (Burlington, VT: Ashgate Publishing, 2012), 73–80; Hannah Scheidt, "The Fleshless Future: A Phenomenological Perspective on Mind Uploading," 315–28, in *Religion and Transhumanism: The Unknown Future of Human Enhancement*, eds. Calvin Mercer and Tracy J. Trothen (Santa Barbara: Praeger, 2015), 315–28; Noreen Herzfeld, "Cybernetic Immortality versus Christian Resurrection," 192–201, in *Resurrection: Theological and Scientific Assessments*, eds. Ted Peters, Robert John Russell, and Michael Walker (Grand Rapids: Eerdmans, 2002); and Ted Peters, *Anticipating Omega: Science, Faith, and Our Ultimate Future*, in the series *Religion, Theology and Natural Science* 7 (Gottingen: Vandenhoeck & Ruprecht, 2006), 130–31, 119.

[18] Much of the material in this chapter is an adaptation from Calvin Mercer's publications, including "Whole Brain Emulation Requires Enhanced Theology, and a 'Handmaiden,'" *Theology and Science* 13, no. 2 (April 2015): 175–86; and "A Theological Assessment of Whole Brain Emulation: On the Path to Superintelligence," *Religion and Human Enhancement: Death, Values, and Morality*, eds. Tracy Trothen and Calvin Mercer, 89–104, in *Palgrave Studies in the Future of Humanity and Its Successors*, series eds. Calvin Mercer and Steve Fuller (New York: Palgrave Macmillan, 2017).

[19] Robert C. Twigg, Thomas Hengen, and Marlyn Bennett, "Going Back to the Roots: Using the Medicine Wheel in the Healing Process," *First Peoples Child & Family Review* 4, no. 1 (November 30, 2008): 10–19.

with new abilities and possibilities. We coined the term "transformational resurrection" for this second kind of raising.

The apostle Paul, the influential early Christian writer who wrote about transformational resurrection, used the term *soma pneumatikon*[20] (literally "body spiritual") to refer to this transformed life. The Christian scriptures indicate that Paul understood that the body, upon death, would be transformed in a similar way to that of the resurrected Jesus Christ, whose bodily resurrection allowed him to move through closed doors[21] and miraculously appear on a remote road.[22] As with the resurrected Christ, the transformation converts the person's perishable,[23] dishonoring,[24] and weak[25] body to one that is imperishable,[26] glorious,[27] and powerful.[28] Importantly, as with Christ, the new life of the one resurrected, while dramatically changed for the age to come, is a clear continuation of the pre-resurrection person.[29]

An objection to interpreting mind uploading as transformational resurrection is that eternal (i.e., resurrected) life and indefinite existence in a digital substrate are different in that resurrection is to eternal life and mind uploading is a continuation of this life. That is true. However, if people start living for hundreds or thousands of years, then notions of the life to come will surely be reinterpreted. Very possibly, death—as we think of it now—as the dividing line between this life and the life to come could recede to the background.

Karmic religions that postulate a soul that moves from body to body, from life to life, may be better positioned than the monotheistic traditions to appreciate and accept mind uploading as a technological means of reincarnation. It would not be simple, however. These traditions would then need to reinterpret death as somehow being included in the uploading process. If the upload is successful in transferring consciousness, then it is not clear the soul has been reincarnated. Perhaps a successful upload will serve only as a way to enhance the current incarnation of the soul.

Embodied Cognition

The ancient scriptural insistence on body in the monotheistic religions actually finds support in current scientific thinking. Research on cognition has moved away from seeing the rest of the body as peripheral to the brain. Exerts now generally see cognition as intertwined with the body in its dynamic interaction

[20] 1 Corinthians 15:44.
[21] John 20:19.
[22] Mark 16:12; Luke 24:13–32.
[23] 1 Corinthians 15:42.
[24] 1 Corinthians 15:43.
[25] Ibid.
[26] 1 Corinthians 15:42.
[27] 1 Corinthians 15:43.
[28] Ibid.
[29] 1 Corinthians 15:35–44.

with the natural and sentient environment. As we learned in the chapter on "Cryonics," a case can be made for "body memory" that is not confined to the brain.[30]

The operative term is "embodied cognition," what one author called "the missing link between robotics and AI."[31] For over two decades, a human-like, embodied approach to robotics has prevailed. In Japan, a leader in robotics, the human form apparently plays an especially important role in robot development, perhaps due in large part to the notion of unity of the material and spiritual in Japanese religions. Notice the subtitle to an important work in this field: "*Being There: Putting Brain, Body, and World Together Again.*"[32] Andy Clark, the author, makes the point that minds are for doing, and, so, require a body.

Embodied cognition raises the question of how can the mind be distinguished from the body in order to be uploaded? To put it another way, how can the memory and personality in the brain be separated from the rest of the body? There are 80–90 billion neurons in the brain, all constantly changing and many of which are critically connected to other neurons outside the brain. The buzz of activity in the brain entails input-output interaction with processes ongoing in the rest of the body. The 100 trillion bacteria living in our gut can cause depression and anxiety. With neurons, chemical transmitters, and microbiotic entities, we are a surging hormonal package, providing a basis for the argument that that the mind, or self, is inseparable from the flesh and blood body.[33]

If our mind or self is not reducible to memory and personality, stored as information only in the brain, then a mind upload, leaving the body behind, would not accomplish the purpose of moving the whole person into a new host. Whole brain emulation could leave behind significant dimensions of who we are. Musicians, artists, and athletes often describe the experience of their bodies knowing what to do, without engagement of the thinking process. A type of automatic pilot sets in, animated by the hands, legs, arms, or other aspects of the embodied self. The brain may not be the sole center of thought.

One might, in theory, envision that the new technological body could be outfitted with all the necessary chemical connector processes. In this way, the

[30] John Swinton, "What the Body Remembers: Theological Reflections on Dementia," *Journal of Religion, Spirituality & Aging* 26 (2014): 160–172.

[31] Anat Ringel Raveh and Boaz Tamir, "From Homo Sapiens to Robo Sapiens: The Evolution of Intelligence," 197–215, in *AI and the Singularity: A Fallacy or Great Opportunity*? eds. Robert K. Logan and Adriana Braga (Basel: MDPI, 2020; originally published in *Information* [December 21, 2018]) 210. The authors go on to argue that embodied cognition "…should not be thought of as an imposition on AI but as a new challenge," p. 210.

[32] Andy Clark, *Being There: Putting Brain, Body, and World Together Again* (Cambridge, MA: MIT Press, 1997). From an explicitly theological perspective, see Victoria Lorrimar, "Mind Uploading and Embodied Cognition: A Theological Response," *Zygon* 54, no. 1 (3/19): 191–206.

[33] Noreen Herzfeld makes this general point well in "Must We Die? Transhumanism, Religion, and the Fear of Death," in *Religion and Human Enhancement: Death, Values, and Morality*, eds. Tracy Trothen and Calvin Mercer, in *Palgrave Studies in the Future of Humanity and Its Successors*, series eds. Calvin Mercer and Steve Fuller (New York: Palgrave Macmillan, 2017), 290–91.

mind upload could be embodied in a behavior-based robot that interacts with its environment much like humans do now. This does, most definitely, complicate things technically. Any simplistic analogy, such as moving a hard drive file to a flash drive, breaks down.

What Kind of Body?

Given the centrality of body in the monotheistic religions, one might conclude that Judaism, Christianity, and Islam would reject mind uploading, because the body, as we usually think of it, is being left behind in the mind uploading procedure. But there are counterarguments. For example, in ancient Christianity, the first-century Corinthians put a question to the Apostle Paul that highlights the issue. "*How* are the dead raised? With what *kind of body* do they come?"[34] The emphases in italics are, of course, ours.

Does "body" have to be flesh and blood? That is a critical question for a theological assessment of mind uploading. The early Christians in Corinth wanted to know what "kind" of body they would have in the resurrection life. Clearly, for Paul it is a transformed body that has continuity with but is quite different from the biological body. The new life is lived in a body that is, as we have seen, imperishable, glorious, and powerful. In fact, Paul explicitly says that "Flesh and blood cannot inherit the kingdom of God … we shall all be changed."[35] Here is how one author worded his rhetorical question:

> … why does a biological body take priority over an artificial body, when both are made up of different combinations of elements on the periodic table?[36]

We have made the point in other contexts that technology for one kind of human enhancement (e.g., whole brain emulation) is not likely to develop in isolation from additional enhancing technologies. By the time, whenever it is, mind uploading capability is here, other related developments, such as robotics and tissue engineering, will surely have matured. Ray Kurzweil was introduced in an earlier chapter and is the most well-known transhumanist enthusiast and advocate. His vision, referred to in a previous chapter, is well-known and merits repeating in this context:

> By the time we have the tools to capture and re-create a human brain with all of its subtleties, we will have plenty of options for twenty-first-century bodies for both nonbiological humans and biological humans who avail themselves of extensions to our intelligence. The human body version 2.0 will include virtual bodies

[34] 1 Corinthians 15:35.

[35] 1 Corinthians 15: 50–51.

[36] Matthew Zaro Fisher, "More Human Than the Human? Toward a 'Transhumanist' Christian Theological Anthropology," in *Religion and Transhumanism: The Unknown Future of Human Enhancement*, eds. Calvin Mercer and Tracy Trothen (Westport, CT: Praeger, 2015), 29.

in completely realistic virtual environments, nanotechnology-based physical bodies, and more.[37]

The often-ignored aesthetic dimensions of transhumanism have been well presented by American designer and artist, and transhumanist, Natasha Vita-More.[38] For example, in a project named "Primo Posthuman," she designed a future body prototype and imagined, through art, how that might look.[39] She is currently Executive Director of the leading transhumanist organization, humanity+, discussed in the chapter, "Transhumanism, the Posthuman, and the Religions."

The theological implications of the emphasis on embodiment in the cognitive sciences are considerable. A mind would not necessarily be uploaded into a cold, faceless piece of hardware. Embodiment does not have to be biological, but even a biological scenario is not unrealistic. While the field of tissue regeneration is focused on therapeutic goals, it could play a role in enhancing the robotic target platform for a mind upload. Even with current capability, the machine hosting the emulation could have a natural skin-like surface and robotic arms, much like we saw in the fictional movie, *Avatar*.

Robots are likely to play increasingly interactive roles in our lives, as the artificial intelligence (AI) field gives more attention to interaction with the environment. An uploaded mind could take advantage of all the advances in robotics to construct alternative body platforms. Although beyond the scope of this chapter, the question of virtual bodies and their relationship to the important embodiment discussion, will find its way into this ongoing conversation.

Allopathic medicine has long been taking us down a path that goes beyond flesh and blood. Many of us are already cyborgs with pacemakers, artificial knees, metal pins and plates, and cochlear implants. It is a bigger leap, for sure, but perhaps a mind uploaded into a robot body could be one way of expressing the transformation Paul envisions. Whether or not that proves to be a case is likely to be the subject of much debate among theologians.

[37] Ray Kurzweil, *The Singularity Is Near: When Humans Transcend Biology* (New York: Viking, 2005), 199. See especially Chaps. 5 and 6.

[38] See, e.g., Natasha Vita-More, "Aesthetics: Bringing the Arts and Design into the Discussion of Transhumanism," in *The Transhumanist Reader: Classical and Contemporary Essays on the Science, Technology, and Philosophy of the Human Future*, 18–27, eds. More, Max and Natasha Vita-More (West Sussex, UK: Wiley-Blackwell, 2013); and "Life Expansion Media," in *The Transhumanist Reader: Classical and Contemporary Essays on the Science, Technology, and Philosophy of the Human Future*, 73–82, eds. More, Max and Natasha Vita-More (West Sussex, UK: Wiley-Blackwell, 2013).

[39] Natasha Vita-More, "The New [human] Genre—Primo Posthuman," Presentation at Ciber@RT Conference (Bilboa, Spain: 2004). http://www.natasha.cc/paper.htm. See also Vita-More, "The Posthuman Future—Interview with Natasha Vita-More," *Studio 360* (November 4, 2011). http://www.wnyc.org/story/233794-posthuman-future/; and Natasha Vita-More, "Design of Life Expansion and the Human Mind," in *Intelligence Unbound: The Future of Uploaded and Machine Minds*, 240–47, eds. Russell Blackford and Damien Broderick (West Sussex, UK: Wiley Blackwell, 2014), 17.

Personal Identity

Even if we can conceptualize mind uploading in a way that meets the religion's scriptural and theological criterion for a body, we are still not there yet. Here we jump into the complicated topic of personal identity and turn for help to the discipline of philosophy.[40] Theology often address topics that are, fundamentally, philosophical issues, and at times it can help to turn to philosophy for input. We do not want to get too involved in the thorny philosophical issue of personal identity. It is, however, important to lay out the basic problem. This will likely be one of the most difficult parts of the textbook to understand.

Honda Accord, Or Not

One problem in personal identity is illustrated in the ancient Theseus' Paradox, which we have encountered earlier in the textbook. Articulated by Plutarch in the first century, the question is whether a ship, in this case the ship of Theseus, where every part has been replaced over time, is still the same ship. We can modernize the illustration by thinking about a particular vehicle today. We use a Honda Accord vehicle as our example. Over the years, the owner has replaced many parts of the Accord. Of course, if the owner replaced every part at the same time (we are not sure what this would look like down at the mechanic shop), then the Accord is being replaced by another Accord.

However, gradual replacement over a period of years of all the parts complicates the question. Even if gradually over the next decade the owner replaces every single part of the Honda Accord, the owner still considers it the same Honda Accord. Over time, every atom of the matter that makes up our bodies is replaced. So, like the ship of Theseus and the Honda Accord, we are not constituted by the same material as several years previously. However, we—and the law—consider ourselves to be the same personal identity as at birth.

[40] An excellent discussion of whole brain emulation, from a respected philosopher who is favorable to the possibility of uploading consciousness, is David J. Chalmers, "Uploading: A Philosophical Analysis," 102–19, in Russell Blackford and Damien Broderick, eds. *Intelligence Unbound: The Future of Uploaded and Machine Minds* (West Sussex, UK: Wiley Blackwell, 2014). Careful critiques of Chalmers are provided by Massimo Pigliucci, "Mind Uploading: A Philosophical Counter-Analysis," 119–31, in *Intelligence Unbound: The Future of Uploaded and Machine Minds*, eds. Russell Blackford and Damien Broderick (West Sussex, UK: Wiley Blackwell, 2014), arguing that consciousness and self-identity are biological phenomena; and Joseph Corabi and Susan Schneider, "If You Upload, Will You Survive?" 131–45, in *Intelligence Unbound: The Future of Uploaded and Machine Minds*, eds. Russell Blackford and Damien Broderick (West Sussex, UK: Wiley Blackwell, 2014).

The Same Thing

A related philosophical distinction that can be important as we move forward in this conversation is the different ways the word "same" is used. There is a difference between (1) Some thing being qualitatively the same, i.e., identical, to some other thing, and (2) Some thing being numerically identical, i.e., the same thing. In case (1), "identical" is being used as meaning "exactly similar." In (2), "identical" does not mean "exactly similar." In this strict philosophical sense, identity is the relationship a thing has with itself.[41]

There are several philosophical theories of personal identity. The leading theory is the mental states theory, sometimes called the psychological theory of personal identity. It goes back to the seventeenth century philosopher John Locke and his influential *Essay Concerning Human Understanding*.[42]

Locke's theory, as amended by current philosophers, is that personal identity continuity entails overlapping or connected chains of mental states, such as memories, beliefs, intentions, desires, and character traits. So, the elements of our body may change over the years, but there is a psychological/mental continuity over time that makes the authors writing this textbook, and you the readers reading this textbook, the same persons that we all were years ago at birth.

The philosophy of personal identity is relevant, because this psychological theory of identity may appear to support the continuation of who we are into an upload. The theory's focus is not on body but on memory or mental states that can, at least theoretically, be copied and uploaded into a computer. Unfortunately, this is not the end of the story.

Locke has a problem and so do those who think who we are can be uploaded into a computer. It is called the "duplication objection." A copy of something, whether it is a sheet of paper sliding off the photocopier glass, or a mind, is not the same numerically identical thing as the original. It is a copy. A copy of the self (or mind), no matter how perfect, cannot be numerically identical to that self. Using the example we introduced above, we take a perfect mold of every single part of the Honda Accord, including the scratches on the passenger door, and make an exact copy. It is not the same, i.e., numerically identical, Honda Accord. It is still a copy.

We can drive home the point with an existential example. Here is how Professor Calvin Mercer, one of your authors, put it in an article on whole brain emulation:

> Logic guides us here, on the principle that two things, different from one another, cannot be identical to the same thing. Logic also seems to confirm intuition. If technology achieves a silicon copy of your brain, will you be comfortable that the copy is you, even if the copy contains the information or information patterns that give rise to your memory, beliefs, ambitions, feelings, intentions, and

[41] Chalmers, "Uploading: A Philosophical Analysis," 108.

[42] See *Essay Concerning Human Understanding* (1689), especially Book II, chapter 27. http://www.earlymoderntexts.com/assets/pdfs/locke1690book2.pdf.

> personality? Put more grossly, imagine you are suffering from an incurable and painful disease and your doctor produces a copy of your brain and says she is ready to give the original you a lethal injection, because your memories, beliefs, and character traits are in the digital substrate ready to continue living. Will you ask for the needle?[43]

The problem comes into even greater focus when we think about making several copies. If we develop the ability to copy a mind once, presumably we can copy it twice or ten times. We do not have ten "yous" out there, because a copy is not the same (numerically identical) as the original. However, we have ten copies that are *qualitatively*—but not *numerically*—the same as the original you. This raises all kinds of theological, ethical, and legal questions, as well as the philosophical one. What is the relationship of the various copies to one another and to the original? This duplication argument seems fatal to the idea that personal identity is preserved in an upload. But, there may be a way around this duplication objection.

Survival, Not Identity

We here peel off another layer of this complicated philosophical issue, which has religious implications. Perhaps what matters is not the continuation of personal identity, but, rather, survival. Reputable philosophers such as Derek Parfit[44] have argued that psychological continuity provides survival, and this is what matters, even if what is surviving is not the exact same (i.e., numerically identical) person. It may seem as if we are just playing with words here, but the survival theory does address the serious duplication objection.

If personal identity is not continued in the mind upload, then we need to address the nature and status of the upload, or uploads. Theologically, from the monotheistic religions, perhaps the "image of God" is flexible enough to include persons who are qualitatively the same as the person whose brain information was copied. It is uncertain if the soul, in the karmic religions, would be interpreted with enough flexibility to fit the "survival, not identity" model.

As we did in the cryonics chapter, let us turn to the Christian doctrine of resurrection for possible help here. The resurrected person has continuity with the pre-death person, just as the Christian scriptures describe how Christ's resurrected body exhibited holes where nails were driven in at crucifixion.

[43] Calvin Mercer, "Whole Brain Emulation Requires Enhanced Theology, and a 'Handmaiden.'" *Theology and Science* 13, no. 2 (April 2015): 181–82.

[44] For Parfit's work on survival, see the important Part III, "Personal Identity," 199–47, in Derek Parfit, *Reasons and Persons* (Oxford: Clarendon Press, 1984). In his discussion of personal identity, Chalmers is influenced by Parfit. See Chalmers "Mind Uploading: A Philosophical Counter-Analysis,"108–14. On Parfit's relevance, see Naomi Wellington, Whole Brain Emulation: Invasive vs. Non-Invasive Methods," 178–92, in *Intelligence Unbound: The Future of Uploaded and Machine Minds*, eds. Russell Blackford and Damien Broderick (West Sussex, UK: Wiley Blackwell, 2014), especially 189–90.

Christ, however, was changed, transformed after death. Resuscitation of a dead body brings the exact same person back to life. Transformational resurrection brings into existence a new being, transformed but also consistent with their past life trajectory. "We shall all be changed," the early Christian writer, apostle Paul, said.[45] To use the philosophical term we have introduced, the person "survives" after death.

Of course, Paul's culturally conditioned reflections are not necessarily directly relevant for the world of nanotechnology and uploading. Paul's language is imaginative, interesting, and potentially supportive of new and creative ways to theologically envision the body in the approaching world of enhancement technology.

In the scenario being discussed, the goals of both transformational resurrection and mind uploading are not necessarily to move the exact (i.e., numerically identical) same person into the new existence. Rather, it is to allow the person to survive and even flourish as a transformed person who has continuity with the same person in the past. In this line of thinking, transformational resurrection can apply to one upload and perhaps provide an example of how at least one of the monotheistic religions could embrace mind uploading.

But, once again, we are not at the end of the story. Transformational resurrection may work for one upload, but the doctrine has to be stretched to accommodate transformation into more than one upload. Perhaps high-fidelity uploads could be viewed in the same way as identical twins. Here, "identical" is being used as qualitatively the same, not numerically the same. Uploads and identical twins are new, different persons.

The philosophical and theological issues at play, and the conclusions people make about mind uploading, will likely be greatly informed by what/who actually results from the process. A battery of psychometric tests administered to a mind upload could be revealing, providing detailed information about cognitive, emotional, and personality status. Unfortunately, we will not have that data to consider unless or until we get a successful mind upload.

In this section on personal identity, we have concentrated on the leading philosophical theory, i.e., Locke's mental states or psychological theory. A second theory, spatiotemporal continuity, is that personal identity is preserved in a body. This theory is not that helpful for mind uploading, because emulation, by definition, entails mind transfer from a body. Also, the spaciotemporal continuity theory is not consistent with the notion of the psychosomatic unity, which we identified as the best scholarly understanding of biblical anthropology.

The narrative theory of personal identity is the newest philosophical theory and one not yet been considered with regard to whole brain emulation. In this newest theory, our identity is generated over time though the evolving story of our past, present, and anticipated future. Supported by developmental,

[45] 1 Corinthians 15:50–51.

personality, and social psychology research, this theory may prove a promising way to think about whole brain emulation.[46]

How the Religions Might Respond

Even if philosophical, theological, and ethical (discussed in the next section) concerns seem overwhelming, none of these concerns will halt scientific work on mind uploading. As long as the technical barriers can be overcome, and Nick Bostrom and many other experts think they can be overcome, then mind uploading in some form will probably transpire sometime this century, polled experts say. So, it is incumbent upon the religions, and society at large, to think through this development before it arrives.

The new world of radical human enhancement, as we have noted, is likely to throw traditional political and religious alignments into atypical patterns. Let us reflect on the many ways this might unfold with regard to mind uploading. As we have detailed in the chapter, "Transhumanism, the Posthuman, and the Religions," conservatives are generally suspicious of science, at least compared with moderates and liberals. On this basis, conservatives, who tend toward the traditional, would be leery and perhaps apprehensive about a procedure as radical as whole brain emulation. Many liberal religionists are likely to join with conservatives in this opposition, although the liberals would likely be motivated by justice concerns.

Further supporting the scenario wherein conservatives reject mind uploading, in a psychological study of fundamentalism,[47] Professor Mercer argued that fundamentalism is primarily a reaction to the perceived loss of identity. Christian fundamentalists exhibit anxiety about their loss of identity as a child of God who will live forever with God in heaven. Mind uploading may very well preserve personal identity, as we have discussed, but that is by no means certain. Hence, normal and understandable anxiety will likely attend advancements that lead to uploading, at least unless and until people are convinced it preserves personal identity.

On the other hand, Christian conservatives and those who believe in a distinct soul, such as Hindus, tend to prefer a dualistic model of human beings as consisting of a body and a soul, with the true self residing in the soul alone. In this otherworldly view, the soul may be understood as the primary object of God's salvific action in Christianity or subject to the law of karma in Hinduism.

Those traditions (conservative Christian and karmic) with notions of the soul/self as distinguishable from the body, could prove to be more compatible with and embracing of mind uploading, which the religious traditions would

[46] A good summary of the theory is found in David Shoemaker, "Personal Identity and Ethics," section 2.3. http://plato.stanford.edu/archives/spr2014/entries/identity-ethics/.

[47] Calvin Mercer, *Slaves to Faith: A Therapist Looks Inside the Fundamentalist Mind* (Westport, CT: Praeger, 2009).

interpret as moving the soul into a new body or, in this case, a computer platform.[48] Mind uploading is one path to superlongevity and could offer the Christian conservative an opportunity to continue living until they are confident they have achieved the required status to go to heaven, and avoid hell, when they die.

As we have seen, many uncertainties surround whole brain emulation. When or if uploading procedures are developed, the end result is unlikely to be merely a copy of the original mind, soul, or person, however faithful the emulation. Since the upload will be placed into a powerful computer, the opportunity, perhaps likelihood, is that the upload will develop into superintelligence. This possibility segues us into the very interesting topic of the next chapter.

Ethical Issues

As we have seen, whole brain emulation cracks open a host of confusing—and interesting—theological issues. Suppose all of them are successfully addressed, and religions that engage mind uploading are theologically comfortable with this enhancement technology. Solving the theological questions does not mean the faith communities would be supportive of developing such technology. Ethical concerns must be addressed.

As with other radical enhancement technology, reasons can be marshalled for both a precautionary and proactionary stance on mind uploading. Should mind uploading work out, benefits of the procedure include the preservation of life and reduction of suffering. On the precautionary side, we will outline challenges facing this radical enhancement. As you read this next section, we invite you to consider if you would advise a primarily precautionary or a primarily proactionary approach to mind uploading, and why.

The Therapy: Enhancement Continuum—What It Means to Make Us Better

We have seen that interventions on the enhancement side of the continuum are generally more controversial than those clearly on the therapy side. Critics will likely see whole brain emulation as pushing the limits of the enhancement side of the continuum. To many people the whole project just seems an outlandish venture more suitable for a science fiction movie. Indeed, the theme has inspired a long list of fiction, films, video games, and even comics.

The target platform for the mind upload is intended to be more reliable and sustainable than the human body. Indeed, that is a main purpose for whole brain emulation. So, an obvious case can be made that whole brain emulation

[48] See Robert M. Geraci, Apocalyptic AI: Visions of Heaven in Robotics, Artificial Intelligence, and Virtual Reality (Oxford, UK: Oxford University Press, 2010), who finds a striking similarity between conservative apocalyptic traditions of Judaism and Christianity and mind uploading advocates, all of whom envision being released to immortality in a glorified new body.

is enhancement, taking the self's capabilities far beyond what is normal. But from a religiously informed perspective, it is important to recall that the dividing point of what can be considered acceptable and unacceptable on this continuum is not the concept of "normal," but the point at which the changed person no longer has religious integrity.

If an uploaded mind is seen as far removed from one's current divinely inspired state, then it may be that we have strayed too far from the state that is understood to be karmically merited or God-intended. How the religions decide what too far is, will be challenging. If one's basic identity would be continuous as an uploaded mind, and if aging and death are regarded as diseases that keep the self or soul from enjoying life, then whole brain emulation may be acceptable to religion. If, however, we leave behind or cut out important parts of ourselves, or perpetuate injustice, then religion would likely reject mind uploading. The engagement of the therapy—enhancement lens by religions is one way to help us explore what makes us authentically better.

As noted earlier in this chapter, much technical improvement is required before a mind upload could be successfully performed. For the foreseeable future, whole brain emulation is likely to be viewed as a radical enhancement and one the religions will judge as being too far removed from the life God intended or the state of the soul that is karmically merited. Religions, however, have proven themselves flexible over the long haul of history, and widespread availability of whole brain emulation would likely result in some degree of theological embrace.

Choice

Choosing to be uploaded, as with other anticipated enhancement options, will affect others as well as ourselves. As with medical procedures, hopefully by the time, if ever, mind uploading is commercially available, the medical risks and dangerous side effects will be known and minimized. If the chance of serious hazards is significant, that information would have to be carefully considered by the prospective consumer and their family and advisors before consent to the procedure.

The choice made will potentially impact not only the patient and people in their life. If the upload is successful, a range of additional impacts have to be considered. That range includes so many possibilities, many unknown, that we can only begin to point at some of the ethical challenges of the choice made.

A primary motivation for developing and choosing mind uploading is to preserve personal identity and consciousness in a more reliable and lasting platform then the human body. This vision is far down the road, of course, but maybe a person will be able to utilize future technologies to fashion a new host for their mind that, ideally, can be tailored to the wishes and needs of the source mind. In such a case, the individual will be enhanced in any number of ways, a reality that can cut two ways. Family and friends will have to adjust to the new presentation and may not like that the person they had previously known is no longer embodied in the same way. Aspects of a relationship, such as physical touch, will be changed. On the other hand, a longer and enhanced life may contribute to making the relationships better. So much is unknown about how mind uploading will turn out.

If the uploading process does not destroy the "source" self, i.e., the consumer or patient choosing the procedure, then the choice to upload becomes a decision to bring a second and new being into existence. Choosing to birth a child brings with it some level of risk that the child may have, for example, a serious debilitating birth defect. Such risks are typically low enough, and the benefits large enough, that prospective parents choose to start families. Mind uploading, at least in the beginning years, may carry a much higher risk in terms of probability for an unhappy outcome. On this point, the ethical calculation may shift as the procedure is improved over time. Elsewhere in this chapter, we raised the possibility of more than one upload resulting from the procedure. That outcome further complicates the choice, because now the source mind, the consumer, is bringing into existence several new beings.

Many, if not all, of the choices about mind uploading will reflect the values of the people making these choices. Recall the top five values you identified as part of your work in the "Radical Human Enhancement and Ethics" chapter, and you can begin to identify the kinds of choices you might make, should this radical option come to pass. For people of faith, values derived from their religious training and commitments will play a major role.

Justice

Whole brain emulation would generate an abundance of ethical and moral questions regarding the status of the one or more uploads.[49] Undoubtedly, depending on the particular outcome, there will be questions regarding uploads about legal status, rights, and responsibilities. The relationship between the upload and the source mind will involve questions about property rights, marital considerations, child-care obligations, and much more.

As with all radical enhancements, distributive justice is a legitimate concern. If mind uploading brings benefits and becomes a desired procedure, there is no guarantee it will be made available in a fair and equitable manner. Concerns about safety raised with other enhancements are also relevant to mind uploading.

Race, gender, size, disability, and some other aspects of our embodied identities have been the basis for systemic discrimination. As discussed in the "Cryonics" chapter, one theoretical way to address body-associated prejudices is to rid ourselves of bodies as traditionally understood. No more flesh and blood bodies, no more body-based discrimination. Mind uploading as

[49] For a sampling of what would be a flood of legal discussions, see Kamil Muzyka, "The Outline of Personhood Law Regarding Artificial Intelligences and Emulated Human Entities, " *Journal of Artificial General Intelligence* 4, no. 3 (December 2013): 164–169. https://doi.org/10.2478/jagi-2013-0010. ISSN 1946-0163.

conceived by transhumanists certainly offers this possibility. In the chapter on "Superlongevity," in the context of the discussion on cyborgs, we considered the idea that this thinking is simplistic. The psychological, ideological, and attitudinal bases of prejudice run deeper than the structural platform (e.g., the new body) that houses our mind.

In fact, mind uploading could heighten the problem of discrimination if partial or defective mind uploads do not meet prejudicial normative standards of society. We might value rational thinking over emotional intelligence or creativity. The religious values of morality, compassion, and justice could get lost in the shuffle. Again, we see the importance of the principle of co-design that ensures diverse and equitable representation at the table as whole brain emulation is developed.

In the chapter, "Radical Human Enhancement and Ethics," we saw the importance of community. Human beings are communal creatures, and all the religions give emphasis to this dimension of our lives. How mind uploading will unfold and what the process will produce is speculative, but how people relate to uploads and how the emulations relate to each other will be important.

Religious ethicists will have plenty of work addressing the once science fiction, now very real, concern that an uploading procedure would not destroy the original (i.e., source) brain. In relation to the original brain, it will be necessary to determine the status of the uploaded copy to the original and the uploaded copies to each other. How this would impact religious institutions, worshipping communities, and covenantal bonds between religious followers is, of course, uncertain.

The analogy of identical twins could be helpful. Identical twins are, well, identical (qualitatively) until each begins its own unique journey. Like a story with multiple endings, each uploaded mind would evolve its own unique life. The advanced technological world may allow the various uploads to be connected to one another in a significant way, allowing for some sort of corporate self. Or, perhaps the emulations will be seen as constituting a new form of family.

Questions for Discussion

1. Do you think we might develop the technology for mind uploading? Why or why not? If so, when?
2. Would you consider uploading your mind? Why or why not?
3. What do you think is the essential aspect that constitutes who you are?
4. How essential is your body to who you are?
5. Do you think the religions will embrace mind uploading, should it become widely available? Why or why not?
6. How might the experience of community change in a world where mind uploading is common?
7. What if mind uploading could lead to a profound connection among people of diverse generations? How important could that be?

CHAPTER 10

Superintelligence: Bringing on the Singularity

Technology

The Future of Artificial Intelligence

In the *New York Times* bestseller, *Sapiens: A History of Humankind*, Yuval Noah Harari writes:

> The more eastern regions of Asia were populated by *Homo erectus*, "Upright Man," who survived there for close to 2 million years, making it the most durable human species ever. This record is unlikely to be broken even by our own species. It is doubtful whether *Homo sapiens* will still be around a thousand years from now, so 2 million years is really out of our league.[1]

While not focused entirely on human enhancement, the author does provide a chapter at the end where he addresses transhumanism. For better or worse, Harari may very well be correct in his prediction that the human species will disappear in time, and the reason may be the emergence of superintelligence.

Russian President Vladimir Putin, in a widely circulated quote, said the country that leads in the development of artificial intelligence (AI) will rule the world:

> Artificial intelligence is the future, not only for Russia, but for all humankind. It comes with colossal opportunities, but also threats that are difficult to predict. Whoever becomes the leader in this sphere will become the ruler of the world.[2]

[1] *Sapiens: A History of Humankind* (New York: Harper Perennial, 2014), 17.

[2] James Vincent, "Putin Says the Nation that Leads in AI 'Will be the Ruler of the World,'" *The Verge* (September 4, 2017). https://www.theverge.com/2017/9/4/16251226/russia-ai-putin-rule-the-world.

C. Mercer, T. J. Trothen, *Religion and the Technological Future*,
https://doi.org/10.1007/978-3-030-62359-3_10

We extend that statement with this statement of our own:

> The country, or company, that develops superintelligence will rule the world for a very short time … and then that superintelligent machine will rule the world—and all of humanity, if the human species is allowed to survive.

That is sensationally put, but it may not be hyperbole.

AI is increasingly becoming a part of our lives. Self-driving vehicles are just one example of AI that is present and quickly accelerating. AI robots navigating the highways do not fall asleep, drive while impaired, or check their texts. AI is used in military simulations, athletic training, medical delivery and diagnostic programs, predictive clinical analytics (e.g., the determination of when might a patient relapse), drug manufacturing research, banking, stock investing, video games, sequencing genomes, diagnosing cardiac conditions, and gene editing, to name just some of the complex functions. Here is an example of AI in healthcare. Microsoft's *Healthbot* is designed to provide the first stage in patient triage. People with health concerns first talk with an intelligent computer agent, which decides if the issue warrants consultation with a human nurse or doctor.

China understands the power in AI and has made huge investments in becoming a world leader. Some of the biggest companies in the world, the so-called "FANG" stocks (Facebook, Apple, Netflix, Google), are all about high tech, with AI central to their products.

We have entered the world of AI, and we are not going back. We cannot go back, because we are addicted. Douglas Estes, drawing upon Yuval Harari's best-selling *Sapiens*, makes the case that, indeed, luxuries become necessities.[3] Lawn mowers, electric lights, and automobiles, when they first appear, are luxury technologies that fast become necessities to which we are addicted. Very few outside of some Amish, Hutterites, and old order Mennonites are willing, or for that matter even able, to transport themselves via horse and buggy. The increased sophistication and pervasiveness of AI may provide a fertile context for the eventual development of superintelligence.

AI on Steroids—Superintelligence

Up to now, computers have been programmed by humans to do some things very well. These machines have long surpassed the human brain at memory and processing speed but were impotent when asked to distinguish between a raisin muffin and spotted dog. When supercomputers, beginning with Deep Blue in the 1990's, beat human champions at chess, the Chinese game "Go," and "Jeopardy," it made news and the public took notice. We now have AlphaZero with Stockfish 8 as a dark horse in the AI chess world. Building on Jeopardy's

[3] *Braving the Future: Christian Faith in a World of Limitless Tech* (Harrisonburg, VA: Herald Press, 2018).

Watson, we developed *Watson for Genomics* and *Watson for Oncology*. These are examples of "narrow" or "weak" AI, machines that can do one thing, very well.

On the one hand, a computer that can beat a grandmaster in chess is a far cry from general human intelligence, much less going beyond that to a superintelligence. However, having the capability to build a machine that can win at this level (i.e., do one thing very well) may eventually lead to a machine that can perform the full range of human cognitive abilities. "Strong" or "general" AI (or "AGI," Artificial General Intelligence) is machine intelligence that equals general human intelligence.

Up to this point in human history, we have used machines as our tools, extending our ability to act on our environment in particular ways. Machines need humans to activate and guide them. We are fast moving into a vastly different terrain—recursive upgrading machines, i.e., machines that continue to learn without human input. We may be on the verge of autonomous machines that will not need us anymore to program them, to teach them.

Unsurprisingly, Ray Kurzweil, perhaps the most well-known transhumanist, has long argued that strong AI is possible and coming. Notable detractors include the well-respected philosopher John Searle.[4] The challenges are significant, as illustrated by this statement:

> The notion of intelligence that advocates of the technological singularity promote does not take into account the full dimension of human intelligence. They treat artificial intelligence as a figure without a ground. Human intelligence as we will show is not based solely on logical operations and computation, but also includes a long list of other characteristics that are unique to humans, which is the ground that supporters of the Singularity ignore. The list includes curiosity, imagination, intuition, emotions, passion, desires, pleasure, aesthetics, joy, purpose, objectives, goals, telos, values, morality, experience, wisdom, judgment, and even humor.[5]

While experts vary on when human level machine intelligence will arrive, Nick Bostrom provides the results of polls taken of technical experts at academic conferences and in professional organizations:

> … it may be reasonable to believe that human-level machine intelligence has a fairly sizeable chance of being developed by mid-century, and that it has a non-trivial chance of being developed considerably sooner or much later; that it might perhaps fairly soon thereafter result in superintelligence …[6]

[4] Although a bit dated, this collection gives some of the long-standing arguments pro and con. See Jay W. Richards, *Are We Spiritual Machines? Ray Kurzweil* vs. *the Critics of Strong AI* (Seattle: Discovery Institute, 2002). A recent technical collection with articles by both skeptics and advocates is *AI and the Singularity: A Fallacy or Great Opportunity?* eds. Robert K. Logan and Adriana Braga (Basel: MDPI, 2020).

[5] Adriana Braga Robert K. Logan, "The Emperor of Strong AI Has No Clothes: Limits to Artificial Intelligence," 5–25, in *AI and the Singularity*, 5.

[6] Nick Bostrom, *Superintelligence: Paths, Dangers, Strategies* (Oxford University, 2014), 21.

Bostrom's last phrase is critical. The story of AI probably will not end when the level of general human intelligence is reached. Strong AI will, perhaps quickly, move beyond general human cognitive ability to superintelligence. The conversation has shifted from whether a machine smarter than the human inventor could be built, to when and how superintelligence will appear.

Appropriately, Bostrom is reluctant to suggest how close we are to building a superintelligent machine. However, when he cautiously suggests that "It seems somewhat likely that it will happen sometime this century, but we don't know for sure"[7] we have some general framework within which to work. Sometime this century, or even in the next two centuries, is quite soon on a historical scale, and others are saying it will happen well before mid-century.[8]

In a previous chapter, we addressed whole brain emulation, which may be the surest path to superintelligence. Whole brain emulation may be a stepping stone to superintelligence by combining an exceptionally intelligent uploaded human mind with AI. However, traditional AI may be the quickest path to the development of superintelligence. AI and whole brain emulation are but two of five possible paths to superintelligence examined by Bostrom.[9]

As we noted in other chapters, technology is moving forward on many fronts simultaneously. So, if machine intelligence reaches and surpasses general human intelligence, we will also see significant advances in robotics and information technology, just to mention two relevant areas for this discussion.

Imagine a superintelligent computer with twice the cognitive power as the average human being. Connected to the internet, it would have wide access to information. Embodied in a robot, it could act on the information and in the world. The first thing it might do is increase its own intelligence to three times that of the average human being. If it could do that, then maybe it could increase its cognitive powers to ten or 100 times the average human intelligence. What would that mean for us and the world? Ironically, it would take superintelligence to answer that question.

The Singularity

Singularity is the term Ray Kurzweil uses to describe the predicted dramatic, sudden future break in human history when general human intelligence is surpassed. His 652-page 2005 most well-known book is titled *The Singularity is Near: When Humans Transcend Biology*. Although quite old with regard to the fast-changing topic of radical enhancement, it is still worth reading to envisage the scope of the transhumanist vision. Kurzweil's is a grand vision, with the following epochs leading up to the sixth and final epoch, the Singularity: (1)

[7] Ibid., vii.

[8] In this more recent survey of experts, 42 percent predicted 2030 or before and only two percent said it would never happen. See Pawel Sysiak, "When Will the First Machine Become Superintelligent?" *Medium* (April 11, 2016).

[9] Bostrom, *Superintelligence*, 22–51. The others are biological cognition, brain-computer interfaces, and networks and organizations.

physics and chemistry, (2) biology and DNA, (3) brains, (4) technology, and (5) the merger of human technology with human intelligence.[10] Kurzweil describes his vision of the sixth and final epoch when "the universe wakes up."

> In the aftermath of the Singularity, intelligence, derived from its biological origins in human brains and its technological origins in human ingenuity, will begin to saturate the matter and energy in its midst ... the "dumb" matter and mechanisms of the universe will be transformed into exquisitely sublime forms of intelligence, which will constitute the sixth epoch in the evolution of patterns of information. This is the ultimate destiny of the Singularity and the universe.[11]

Kurzweil has been working on his vision, using religious language, for a long time. In 1999, he wrote a book titled *The Age of Spiritual Machines: When Computers Exceed Human Intelligence.*[12]

We have seen that the roots of the transhumanist movement, in the latter half of the twentieth century, reflected a secular, science-based outlook. Kurzweil's vision, with its language of the universe waking up, gives expression to the transhumanist vision clothed in generic religious or spiritual language. Indeed, transhumanism has been interpreted as fulfilling religious impulses. For example,

> In effect then, given transhumanism's apparent grand ambitions to somehow definitively resolve human suffering and limitation using the lens of science and technology, the movement could perhaps offer atheists the kind of all-encompassing, highly compelling philosophical narrative—perhaps fulfilling an innate human need that those of a strongly atheistic disposition aren't able to otherwise satiate via religion.[13]

The Singularity could bring extraordinary benefits, such as increased abilities to solve seemingly intractable problems, and it could also bring apocalyptic dangers. Here is Kurzweil's description of the Singularity. We have also included, in the last two sentences, Kurzweil's statement about the existential import of the Singularity. Envisioning the Singularity can profoundly alter one's life vision, religious and otherwise.

> It's a future period during which the pace of technological change will be so rapid, its impact so deep, that human life will be irreversibly transformed. Although neither utopian nor dystopian, this epoch will transform the concepts

[10] Ray Kurzweil, *The Singularity Is Near: When Humans Transcend Biology* (New York: Viking, 2005), 14–21.

[11] Ibid, 21.

[12] New York: Viking Penguin, 1999.

[13] James Michael MacFarlane, *Transhumanism as a New Social Movement: The Techno-Centred Imagination*, in *Palgrave Studies in the Future of Humanity and its Successors*, eds. Calvin Mercer and Steve Fuller (New York: Palgrave Macmillan, 2020), 183. See the discussion of "Transhumanism as a Quasi-religious Movement?," pp. 185–192.

that we rely on to give meaning to our lives, from our business models to the cycle of human life, including death itself. Understanding the Singularity will alter our perspective on the significance of our past and the ramifications for our future. To truly understand it inherently changes one's view of life in general and one's own particular life.[14]

Here is a dramatic definition from Kevin Kelly, the founder of *Wired Magazine*, which focuses on emerging technologies, such as human enhancement technologies. The Singularity is the point at which "all the change in the last million years will be superseded by the change in the next five minutes."[15] The Singularity will not necessarily come at the same time that superintelligence is developed, but superintelligence will most definitely contribute to the Singularity and hasten its arrival.

Bostrom and others are concerned about what happens when superintelligence is developed. With the race on by countries and corporations to achieve this major breakthrough, the worry is that too little attention is being paid to ensuring that the long-term impact of this level of AI is beneficial and not destructive to the planet and humans. We address this concern later in this chapter.

Religious Issues

Building a New Deity in the Computer Lab?

By superintelligence, we are talking about intelligence that surpasses the typical intelligence of an average human being. How far it surpasses will have implications, potentially good and bad, for society. The possibility of superintelligence also has religious implications.[16]

As newer and newer generations of superintelligent machines evolve, with self-replicating capability, they will be able to, theoretically, approach infinite intelligence—"super" intelligence. Theologically, superintelligence is on a path toward a type of omniscience, the word used in monotheistic religions to describe a central attribute of God. *Omni* is the prefix meaning "all" or "every," and *sciens* is Latin for "knowing."

[14] Kurzweil, *The Singularity Is Near*, 7. This definition of singularity is very different than the definition for singularity used in math or physics. See also Ray Kurzweil and Neil Degrasse, "2029 Singularity Year—Neil Degrasse Tyson & Ray Kurzweil," (April 21, 2016). https://www.youtube.com/watch?v=EyFYFjESkWU. Degrasse takes a different view than Kurzweil.

[15] http://webmindset.net/selected-quotes-kevin-kelly/.

[16] Philosophy can provide a helpful context for understanding religious issues. An excellent discussion of superintelligence and singularity by a respected philosopher is David Chalmers, "The Singularity: A Philosophical Analysis," *The Journal of Consciousness Studies* 17 (2010): 7–65. http://consc.net/papers/singularity.pdf. For a critique of Chalmers, see Massimo Pigliucci, "Mind Uploading: A Philosophical Counter-Analysis," 119–31, in *Intelligence Unbound: The Future of Uploaded and Machine Minds,* eds. Russell Blackford and Damien Broderick (West Sussex, UK: Wiley Blackwell, 2014).

Superintelligent machines are not going to exist in a vacuum. Via robotics and possibly other innovations, they will be connected to means to act in the world. Superintelligence is, therefore, going to bring with it super power. The theological word for that is omnipotence, i.e., all powerful.

Omniscience and omnipotence are two characteristics usually attributed to deity, especially in the monotheistic religions. From the karmic religions, such as Hinduism, we can loosely relate pantheism (God is everything and everything is God) to superintelligence that could conceivably express itself everywhere. We have shown how AI, in its weak or narrow form, is already becoming integral to nearly every aspect of our lives. So, superintelligence would not necessarily be confined to one location. It could very well be dispersed, in a way similar to data storage in the "Cloud." Such a dispersal would add to the almost omniscient and omnipresent capacity of superintelligence.

From the perspective of religion, and especially the pantheistic traditions, Kurzweil's vision that the universe will "wake up" is almost theological in that he predicts an AI universe that becomes saturated with potentially divine-like qualities of supreme intelligence and power. Much depends on how intelligence and power are understood, topics explored later in this chapter. Kurzweil's religious background is eclectic, and his description of the universe as waking up is especially interesting, since "awake" is an important term in Buddhism. After his death, Gautama was called "Buddha," Sanskrit for "awakened one."

Soon after his enlightenment, his waking up, the Buddha began wandering around India and was recognized as an extraordinary being. One of the stories, in different versions,[17] goes like this.

> They asked him, "Are you a god?"
> "No," he replied.
> "Are you a reincarnation of god?"
> "No," he replied.
> "Are you a wizard, then?"
> "No."
> "Well, are you a man?"
> "No."
> "So, what are you?" they asked, being very perplexed.
> "I am awake."

From ancient Hindu scriptures, we also see the divine (i.e., Brahman) as omnipresent and, interestingly, the word "awake" is prominent.

> Verily, in the beginning this world was Brahman, the limitless One—limitless to the east, limitless to the north … limitless in every direction … He whose soul is space … In the dissolution of the world He alone remains awake. From that

[17] See, e.g., https://college.uchicago.edu/news/student-stories/i-am-awake and https://teachingsofthebuddha.com/i_am_awake.htm.

> space, He, assuredly, awakes this world, which is a mass of thought. It is thought by Him, and in Him it disappears. His is that shining from which gives heat in yonder sun and which is the brilliant light in a smokeless fire, as also the fire in the stomach which cooks the food. For thus it has been said "He who is in the fire, and he who is here in the heart, and he who is yonder in the sun—he is one."[18]

Maybe Kurzweil's vision reaches too far. Perhaps superintelligence will not bring about Singularity. Modest superintelligence scenarios, short of the Singularity, are possible. Even work toward a superintelligence, an intellect surpassing the human mind but not nearly as far-reaching as the Singularity, can be interpreted as humans reaching for something transcendent beyond our finite and fallible selves. But for sure the vision of a universe waking up, of supreme intelligence and power everywhere, can certainly be understood as a vision not unlike religious visions.

Religion has been interpreted in a variety of naturalistic ways, as originating to meet social, psychological, or economic needs. From these reductionist perspectives, humans created religion and God as well. Superintelligence may be understood as humanity's latest attempt, even if unconsciously, to transcend by trying to create a technological divinity.

Building a Devil in the Computer Lab?

Omnibenevolence, or supreme love, is usually considered a major attribute of God, along with all-knowledge and all-power. Here, we encounter a particularly serious concern. It is generally understood in the monotheistic religions that the knowledge and power of God will be used for beneficial purposes. However, there is no guarantee that a superintelligent machine with extraordinary knowledge and power will act lovingly. Perhaps the outcome will be an omnimalevolent superintelligence.

In an article aptly titled "AI is My Shepherd," the author illuminates the issue:

> If AI is a deity, it's not likely to be the kind that forgives you, showers you with mercy, and sweeps you up in her loving arms.[19]

A friend of one of your authors quipped pessimistically about the above statement, "Well, an AI deity will have initially been made in the image of its maker, humans. So why would we expect it to be any different."

A deity—or machine—with supreme intelligence and supreme power, but not love, does solve the problem of evil and suffering (i.e., theodicy) that we raised in a previous chapter. It is God's love that makes evil and suffering problematic, because an omniscient and omnipotent deity would certainly desire to

[18] *Chandogya Upanishad* 6.12, R. E. Hume, trans., *The Thirteen Principal Upanishads* (London: Oxford University Press, 1934), 247.

[19] Adam Ferriss, "AI is My Shepherd," *Wired* 26, no. 03 (March 2018): 15.

relieve suffering, if that deity were also all-loving. But a machine with powers approaching omniscience and omnipotence, but not omnibenevolence, could conceivably inflict suffering and hardship on an unimaginable scale.

Many religious explanations have been posited for the existence of evil and suffering. We will briefly mention two. Some theologians in the monotheistic religions address the problem of evil and suffering by asserting God chooses to limit divine power in favor of affording humans freedom of choice. In the exercise of choice, humans can do good or evil, benefit the world or destroy it. Human freedom is at the heart of the matter. The worry is that an extremely powerful and knowledgeable superintelligent machine will override human freedom, thwarting the potential for a peaceful coexistence of superintelligence and humanity.

A second solution also works by compromising the power of God. In the chapter, "Transhumanism, the Posthuman, and the Religions," we presented dualistic thinking as informing one of the conceptualizations of the divine. In this model, God's power is limited to some significant degree in the face of the personification of evil in a Satan figure. In other words, the problem of evil and suffering is explained by the work of the devil. The concern about superintelligence, put theologically, is that we may be creating a machine that approaches all knowledge and all power, but without beneficent intent "programmed in." Without successful moral enhancement through loading values and qualities such as empathy into intellectually powerful machines, superintelligence could become more akin to the devil than to God.[20] But, as we see later in this chapter, value loading, if successful, does not necessarily quell the danger. What values and qualities are selected and who does the programming are ethical issues.

Tower of Babel Story

The Tower of Babel story in the Hebrew Bible's book of Genesis provides an interesting lens through which to think about superintelligence.[21] In that story the human community attempts to build a tower to the heavens. In one interpretation, the human builders are trying to "be like God," to borrow a term from the earlier creation story in Genesis.[22] Put theologically, perhaps that is what humans are trying to do with their technology; humans are trying to achieve divine status. In the monotheistic religions, that impulse is regarded as sinful arrogance, incongruent with the will of God. So, in the Tower of Babel story, God thwarts that effort to reach the heavens.

[20] A good discussion of value-loading is Ben Goertzel and Joel Pitt, "Nine Ways to Bias Open-Source Artificial General Intelligence Toward Friendliness," in *Intelligence Unbound: The Future of Uploaded and Machine Minds,* 61–89, eds. Russell Blackford and Damien Broderick (West Sussex, UK: Wiley Blackwell, 2014).

[21] Genesis 11.

[22] Genesis 3:5.

Ideally, superintelligence emerges in a way that, at least for the religions, does not sink into an idolatrous attempt to supplant God, but respects and coexists with humans and exhibits the best of humanity's values. It may be a tall order, especially when superintelligent machines build even more potent machines. By building superintelligence, are humans arrogantly playing God or are humans behaving in creative ways that reflect what the religions believe? How the religions respond to these kinds of questions will help determine not only the shape of future technology but the future of religion.

God as Emergent

Mozart's Symphony No. 41, the famous Jupiter Symphony, emerged from the combined sounds of horns, trumpets, oboes, flute, and strings. In other words, something quite remarkable and sophisticated emerged from simpler random notes and sounds arranged in a certain way. Turning from music to painting, train your gaze to one brush stroke on a Picasso, and it is nothing too remarkable. Pull back the gaze, and it is clear the master coordinated all the brush strokes into great art.

The human brain, as a physical organ of the body, consists of neurons, neurotransmitters, blood glucose, oxygen, electrical impulses, chemical reactions, and a host of other parts and processes. From all this we get sensation, perception, and subjective conscious experience. Something quite profound emerges from the intricate combination and blending of these otherwise seemingly random "brush strokes."

Here is a more formal statement of the theory:

> Emergence theory is the view that new structures, capacities, and processes will come to existence, that these cannot be reduced to the lower level, and that they can exercise a causal influence downwards. So the mental, such as consciousness, is derived from the biological/physical basis but is not to be reduced to it …[23]

One way to think about God is as emergent, or partially emergent, from complexity. The universe, like the human brain in our analogy, encompasses physical parts and processes, although countless times more parts and processes than musical notes or artists' brushstrokes. If the complex processes of the brain can emerge from its material elements and operations, perhaps God emerges and re-emerges, or develops continually, from the cosmos.

Emergence is a nontraditional notion of God and complicates the monotheistic understanding of God as existing prior to the universe and bringing everything into existence through divine creative acts. Emergence theory, however,

[23] Philip Clayton, *Mind and Emergence: From Quantum to Consciousness* (Oxford: Oxford University, 2004), vi. See also Karl Popper, "Natural Selection and the Emergence of Mind," *Dialectica* 32 (1978): 339–55.

is a way of understanding deity that has some support, and dissent, in Christianity.[24]

Emergence as a way of understanding a deity finds some potential support in "process theology," a type of theology introduced in the chapter, "Transhumanism, the Posthuman, and the Religions." As we explained in that chapter, process theology recognizes that change is fundamental to the nature of reality and can be applied to God. God is in process of becoming, like everything else, especially since God is believed to be in relationship with the entire universe and every relationship changes. While process theology was developed by Christian theologians, we saw in that earlier chapter that change and process is an integral notion in karmic religions as well.

Emergence as a way of understanding God need not be viewed as diminishing God's importance and power. One human brain's capability and activity, emerging from a few pounds of flesh, is remarkable. The capability of a deity emerging from an entire cosmos is incalculable. Given the vastness and complexity of the universe, an emergent God is not one to take lightly.

God as emergent provides an interesting lens through which to interpret superintelligence which, theoretically, could become increasingly potent, intelligent, and good or, unfortunately, evil. At the very least, superintelligence with divine-like attributes—but not necessarily being divine—might be understood as a new being, even god-like, emerging from what started out as mundane transistors and electrical signals in your laptop computer.

Dinosaurs Are Extinct—Are Humans Next?

Tyrannosaurus Rex, affectionately known as "T-Rex," and all her cousins are long gone, their place taken over by more adaptable creatures in the sometimes-harsh evolutionary process. Could *Homo sapiens* also become extinct? Apart from any theologically based insistence about how things have to be, the answer is yes, of course, human beings could be replaced and life continue. Do religious visions allow for the possible termination of our species?

Religious and non-religious opinion about human enhancement, that is, the enhancement of—not the elimination of—human beings, is divided. However, when we talk about *post*-human, that is, superseding human beings with super-intelligent machines, then the opinion can quickly shift to resistance.

Before addressing religion specifically, it is important to acknowledge a basic truth overshadowing the discussion in this section. Human beings have a deep-seated and very robust survival instinct.[25] Human self-preservation is not unique, of course. Self-preservation is a primitive trait found in almost all

[24] Joanna Leidenhag, "A Critique of Emergent Theologies," *Zygon: A Journal of Religion & Science* 51, no. 4 (December 2016): 867–882.
https://doi.org/10.1111/zygo.12300.

[25] This and related issues surrounding transhumanism and religion are discussed in Noreen Herzfeld, "Must We Die? Transhumanism, Religion, and the Fear of Death," in *Religion and Human Enhancement: Death, Values, and Morality*, eds. Tracy J. Trothen and Calvin Mercer, in

organisms. According to biologists, this survival instinct is the basis of evolution in higher animals and plants. The individual instinct to self-preservation has many implications, including the formation of social groups, in part because grouping together enhances survival, and grouping also supports the possibility of offspring. It is understandable that our deep psychological need to survive, consciously or unconsciously, might lead some to negatively evaluate transhuman possibilities, like superintelligence, that could spell the end of humanity, or at least human life as we know it.

The religions of the world are generally oriented around the human story. Sacred texts, doctrines, rituals, institutions, ethics, and spirituality are all generated by and addressed to human beings. It is understandable that transhumanist calls for superintelligence or other programs perhaps making obsolete the human race could collide with deep religious impulses, as well as self-preservation instincts.

While true of all religions, the monotheistic ones are especially focused on the human player in the divine drama. In the Hebrew scripture story of creation in Genesis, Chap. 1, to emphasize their elevated status, Adam and Eve are created last in that creation story and, following their creation, God is reported to have declared the created earth creatures "*very* good."[26] In all the days of creation up to this point, the assessment of what was created is "good."[27] So, this creation story emphasizes that the best creation (human) is saved for last, is declared "very good," and allows for the deity to rest, now having completed the perfect universe.

These psychological and religious attachments to the survival of the species, however, do not necessarily mean that human beings are forever and throughout the vast universe to be the sole focus of the religions. As mentioned before, an evolutionary theology understands that our created ancestors were not human but were humanity's predecessors. Using monotheistic language, human beings do not have to be the only object of God's activity toward sentient beings. "Sentience" refers to the ability to perceive, feel, and experience subjectively. The karmic religions, as well, do not necessarily have to focus only on human beings. Indeed, they are better positioned than the monotheistic faiths to shift focus to life forms other than human, because karma and reincarnation applies to all life beings, not just humans.

Religions generally teach humbleness as a virtue or, putting it another way, pride and selfishness are unwelcome sins exhibited possibly most often by those with undue social and political power.[28] Perhaps it is the height of arrogance to think that humans are the only beings made in God's image, the focus of divine

Palgrave Studies in the Future of Humanity and Its Successors, series eds. Calvin Mercer and Steve Fuller (New York: Palgrave Macmillan, 2017), 285–300.

285–299.

[26] Genesis 1:31. Italics are ours.

[27] e.g., Genesis 1:4, 10, 12.

[28] Scholars who take an intersectional, and particularly feminist, approach to religions have shown, too, that not all people suffer from this sin of excessive pride or arrogance. Many of the

attention, and more worthy than any other living being, past or present. Or, maybe it is a lack of creative thinking, fear, or something else that prevents critique of the idea that humans are at the center of God's plans.

If the evolutionary process is used by God, or the divine order, as a method of creation (evolution is widely accepted in the karmic religions, more debated in the monotheistic ones), perhaps that process can be seen as continuing with the development of more advanced species, *techno sapiens* replacing *Homo sapiens*. The idea of humans as created co-creators, discussed in the chapter, "Transhumanism, the Posthuman, and the Religions," is relevant here.

Maybe it could be viewed as a beautiful thing for human beings to be God's created co-creators of a new life-form that brings superintelligence to the planet, along with more love (if we can value-load that in) and less pain and suffering. The ancient Jewish idea of covenant, of God and people in mutual relationship, supports the vision that life is re-created anew and better again and again. The resulting superintelligent beings could, in a sense, be understood as our loving and respectful children who grow up and make us proud. Or, in a less friendly scenario, we become their pets and they our masters. Or, perhaps we worship them as deities of some sort. However it turns out, we may not have a choice in the matter unless we become far more intentional about our values, our beliefs, and the technological future. Even then, it may be out of our hands. Superintelligence, should it arrive, will not fit any category of technology now known.

The religions should be able to articulate a positive reason for humans to be in the world, other than the negative fear of extinction. What, for example, would be made better by humanity's continued existence in an AI world? Your textbook authors are not advocating that we become extinct like the T-Rex. We do think, however, that a question about the indispensability of humanity is a legitimate one to be grappled with by the religions.

Martians and Superintelligence—What Is the Difference?[29]

Based on the prediction of experts reviewed earlier in this chapter, the religions, sooner or later, are probably going to live into a world where superintelligence is a reality. Even if superintelligence is not developed, discussion of the possibility is valuable as a thought experiment, given the increasing capability and prevalence of AI. While this section focuses on a religious assessment of

more socially marginalized err in the direction of not having enough pride, self-love, or awareness of their own power.

[29] Some of the ideas in this and the next two subsections are drawn from Calvin Mercer, "A Theological Embrace of Transhuman and Posthuman Beings," *Perspectives on Science and Christian Faith* 72/2 (June 2020): 1–6; and Calvin Mercer, "A Theological Assessment of Whole Brain Emulation: On the Path to Superintelligence, 95–104, in *Religion and Human Enhancement: Death, Values, and Morality,* eds. Tracy J. Trothen and Calvin Mercer, in *Palgrave Studies in the Future of Humanity and Its Successors*, series eds. Calvin Mercer and Steve Fuller (New York: Palgrave Macmillan, 2017).

superintelligence, an overarching consideration of superintelligence is well articulated by sociologist and bioethicist James J. Hughes, who writes from a transhumanist and Buddhist perspective:

> The most important disagreement between bioLuddites and transhumanists is over who we should grant citizenship, with all its rights and protections. BioLuddites advocate "human-racism," that citizenship and rights have something to do with simply having a human genome. Transhumanists, along with most bioethicists and the Western democratic tradition itself, believe citizenship should be based on "personhood," having feelings and consciousness. The struggle to replace human-racism with personhood can be found at the beginnings and ends of life, and at the imaginary lines between humans and animals, and humans and posthumans.[30]

One way to think about how the religions can, or might, interpret superintelligence is to use the analogy of extraterrestrial life. This analogy is helpful, because the relationship of religion to other populated worlds has long been discussed by theologians and other scholars of religion.[31] How the religions address the possibility of extraterrestrial life could inform how they might address superintelligence, since both may be considered alien.

The possibility of intelligent life beyond Earth has been raised by the vastness of the universe. Our tiny little solar system, with its sun and planets, is hidden in one of the spiraling arms that stretch out from the center of our galaxy, the Milky Way. The Milky Way galaxy has a diameter of 10 to the 18th power. To drive that number home, the diameter of our medium size galaxy would take light 100,000 to 180,000 years to travel. Andromeda, the closest significant galaxy to ours, is twice as large as the Milky Way. The stars, just in our Milky Way, number 400 billion or more, with about ten new stars formed annually. Right now, there are an estimated 30–60 billion galaxies in the universe. Plenty of experts think that there is a significant possibility that intelligent life exists somewhere in this vastness.

The question of "other worlds," i.e., extraterrestrial life, is a theoretical example of divinely created and inspired sentient beings other than *Homo sapiens.* One could mount a religious argument that such life does not exist, that humans are alone in the universe. But that position has not won the day, at least among many scientists and those few scholars of religion who have engaged this question. Ted Peters, a leading Christian theologian in the science and religion field, calls for "Exotheology," which he defines as speculation on

[30] James Hughes, *Citizen Cyborg: Why Democratic Societies Must Respond to the Redesigned Human of the Future* (Boulder, CO: Westview Press, 2004), 75.

[31] One of the best books on this topic is David Wilkinson, *Science, Religion, and the Search for Extraterrestrial Intelligence* (Oxford: Oxford University, 2013).

the theological significance of extraterrestrial life,[32] and expresses concern about "earth chauvinism."[33]

At this point there is no physical evidence of sentient life outside planet Earth. But if intelligent life presented itself to Earth, how would, or should, the religions interpret that life?

Baptizing Aliens

In a widely publicized quote, Pope Francis of the Roman Catholic branch of the Christian religion said the church under his direction would baptize a Martian, should that opportunity present itself.[34]

Most reflection on religion and other worlds has been done by scholars and theologians of Christianity,[35] and that religion provides our main context for considering how intelligent, sentient, extraterrestrial life might be theologically embraced. Possibly, scholars and theologians of Christianity spend more time on this topic than, for example, the karmic religions because Christianity can be seen to be human focused with its doctrine of the incarnation of Jesus Christ. In the karmic religions, karma and reincarnation are more obviously universal. Those traditions place no limits in the universe where karma and reincarnation might apply.

With regard to other worlds, liberal thinking in the Christian religion is much more aligned with a broad karmic outlook than with that of Christians at the more conservative end of the spectrum, especially given the conservative tendency toward literal interpretation of scripture. For conservative Christians, an Earth-centric focus results from a literal interpretation of the creation stories, incarnation, and end-time speculation. Our exploration of how Christianity might embrace alien life is likely more palatable to a liberal and metaphorical interpretation of scripture and doctrine.

[32] *Science, Theology, and Ethics* (Burlington, VT: Ashgate, 2003), in a chapter devoted to the topic and titled "Exotheology: Speculations on Extraterrestrial Life," 121–36. Peters provides a review of the long discussion of this issue in Christian theology.

[33] Ibid., 125. *Theology and Science* 16/4 (2018) provides a theme issue devoted to "Astrotheology & Astroethics." Ted Peters provides a lead editorial.

[34] Abby Ohlheiser, "Pope Francis Says He Would Definitely Baptize Aliens If They Asked Him To," *The Atlantic* (May 12, 2014). https://www.theatlantic.com/international/archive/2014/05/pope-francis-says-he-would-definitely-baptize-aliens-if-they-wanted-it/362106/. See also Edmund Michael Lazzari, "Would St. Thomas Aquinas Baptize an Extraterrestrial?" *New Blackfriars* (2017): 440–57. https://doi.org/10.1111/nbfr.12319. The respected evangelical organization, BioLogos, pondered this question in a panel discussion, "Life Beyond Earth: What Would It Mean for Christians?" See https://biologos.org/resources/life-beyond-earth-what-would-it-mean-for-christians. Stopping far short of answering the question in the subtitle, the panel encouraged exploration of various ways to understand, biblically and theologically, what intelligent life beyond earth would mean.

[35] David A. Weintraub, *Religions and Extraterrestrial Life: How Will We Deal With It?* (New York: Springer, 2014) does a good job marshalling statements from the religions about extraterrestrial life. That most attention has been given to this topic by Christian writers is reflected in that fact that the book has exactly twice as many pages reviewing Christian statements as all the other religions put together.

In one sense, Christ is understood by Christians as particular to planet Earth and the incarnation in Jesus. However, in orthodox teachings Jesus Christ is the incarnation, at a particular space and time, of the eternal "Word" (Greek, *logos*) that transcends space and time: "In the beginning was the Word, and the Word was with God, and the Word was God."[36] The incarnation became a central, for some the central, tenet of Christian doctrine.

According to the doctrine of the incarnation, God came to human beings in a way that continues to speak intimately and effectively to them. Jesus was a Palestinian Jew who spoke Aramaic, wore sandals, and rode a donkey into town. If God, the eternal Word, had chosen to incarnate in twenty-first century Europe, the incarnation would not have taken the same form as in first-century Palestine. For folk living in India, the eternal Word, or however the transcendent or the sacred is imagined, may very well have come in an incarnation that made sense in that cultural and religious context.

Christians might explore the possibility that the eternal Word has been incarnated in other locations in the universe and in ways that speak intimately to other sentient beings, even as Jesus spoke on the hillsides of the Galilee to *Homo sapiens.* The very essence of incarnation, after all, is that God comes to humans (or aliens?) in a form that speaks to humans (or aliens?). Within this framework, a liberal view can interpret anew many biblical texts, such as the statement that God reconciles "all things, whether on earth or in heaven."[37]

Reflections by academic theologians on the implications for Christian theology of extraterrestrial life may not be directly transferable to transhuman/posthuman beings, but such theologizing provides a fertile starting point for the contention that transhuman/posthuman beings are divine creations and can enter into the same kinds of relationships with God that humans can. For example, twentieth century Protestant Christian theologian Paul Tillich concludes:

> Incarnation is unique for the special group in which it happens, but it is not unique in the sense that other singular incarnations for other unique worlds are excluded. Man cannot claim to occupy the only possible place for incarnation.[38]

While we provide the example of Christianity, all religions will likely have to face the question of what doctrinal tenet or religious practice, if any, would be challenged if there are sentient creatures beyond Earth.

Baptizing Superintelligence

We have suggested that Christianity may be the religion that finds it most difficult to embrace intelligent alien beings, in the same way as humans are

[36] John 1:1.
[37] Colossians 1:20.
[38] *Systematic Theology*, Vol. 2 (Chicago: University of Chicago, 1953), 95 f.

embraced. We have shown, however, that, at least on the liberal wing, Christianity could reasonably chart that path, as Pope Francis suggested, should the time come. The same kind of theological reflection could also lead to embrace of superintelligent machines. The karmic religions will probably have even less difficulty making this move.

A superintelligent machine may actually be easier to accept than an alien from the far side of our galaxy. As we have seen, a superintelligent machine could be incarnated, if we want to use that word, in a fleshy robotic body. The cognitive powers could be such that this being would exhibit, for all practical purposes, emotion and would have (or at least will seem to have) experiences that it can report and discuss. The following assessment, pretty typical from roboticists, comes from Scottish AI researcher David Levy, and is several years old:

> We are in sight of the technologies that will endow robots with consciousness, making them as deserving of human-like rights as we are; robots who will be governed by ethical constraints and laws, just as we are; robots who live, and who welcome being loved, and who make love, just as we do; and robots who can reproduce. This is not fantasy—it is how the world will be, as the possibilities of Artificial Intelligence are revealed to be almost without limit."[39]

To get a sense of how tempting it will be (or is) to experience intelligent robots as persons, spend a bit of time googling robots and taking a look at some of them. The Japanese engineers have been especially adept at making robots human-like. As one person put it, they seem "creepily" human. The more pervasive and human-like robots become, the more debates will unfold around robot rights, treatment of robots, the definition of personhood, and a host of religious questions, all to be intensified when superintelligent robots arrive.

Given a world of intelligent, sentient, robots, what is the difference between a Martian and a superintelligent machine? Perhaps Pope Francis would also sanction baptizing these superintelligent machines/beings. Here is a Presbyterian pastor, Rev. Dr. Christopher J. Benek, a leader in the Christian Transhumanist Association, who answers affirmatively:

> I don't see Christ's redemption limited to human beings. It's redemption of all creation, even AI. If AI is autonomous, then we should encourage it to participate in Christ's redemptive purposes in the world.[40]

[39] David Levy, *Robots Unlimited: Life in a Virtual Age* (Wellesley, Mass.: A. K. Peters, 2006): 293.

[40] Hilary Bird, "AI Innovation Could Cause an Ethical Conundrum for Organized Religion," *VB* (October 16, 2017). https://venturebeat.com/2017/10/16/ai-innovation-could-cause-an-ethical-conundrum-for-organized-religion/.

Ethical Issues

Many of the values and principles discussed in the chapter, "Radical Human Enhancement and Ethics," apply to superintelligence, such as its costly development, fair distribution of potential benefits, and implications for community. The precautionary and proactionary distinction is particularly interesting here, because of the potential scale of both the danger and the benefit of superintelligence. Curing a disease or unintentionally spreading a virus now escalates to solving many or all of our problems or destroying humanity and the planet with it. The stakes are breathtakingly enormous.

To ponder the enormity of the stakes, consider the challenge of global climate change. The scientific consensus is that global warming is an impending disaster for the planet and its inhabitants. Also, we know that the crisis is human-made. At this point it is uncertain if sufficient political will exists in enough countries around the world, and especially in countries causing the most damage, to act in time to avert an irreversible catastrophe.

While many helpful "small" technological innovations (e.g., solar power, long-life batteries, energy efficient machines) are being developed, there is no one big technological solution on the horizon. Would a superintelligence put climate damage into overdrive? Could a superintelligence 1000 (or more) times the cognitive power of the average human come up with some technological fix that thus far eludes our relatively feeble human brains?

The Therapy—Enhancement Continuum: What it Means to Make us Better

If we define the issue of superintelligence primarily as a therapy–enhancement issue, the main question for the religions is whether followers could retain or even deepen (truly enhance) their religious integrity with superintelligence. By religious integrity, as we have said earlier in this textbook, we mean being true to how one understands the core values and beliefs of one's religion. In sum, the therapy—enhancement continuum spans technology that is clearly therapeutic and adds little beyond restoring one's health, to technology that is clearly enhancing with no reparative value. Whether an intervention has reparative value depends, of course, on what one considers to be a disease or other condition that is unnecessary to being human. If one considers aging to be a disease and not a normal and acceptable aspect of being human, then anti-aging or prolongevity interventions would be considered therapeutic and enhancing.

Technological interventions falling anywhere along this continuum may be considered acceptable, from a religious perspective, if they are consistent with one's religious commitments and identity. However, if the intervention compromises one's religious integrity, then it will likely be considered unacceptable from a religious perspective. Technological interventions that fall in the clearly therapeutic zone will tend to be more religiously acceptable since these interventions are medically healing and/or minimize suffering, both of which are usually consistent with religious teachings. Regarding superintelligence, much depends on the capacity and inclination of superintelligence to do good and to help us to realize the goals for good lives, and potentially afterlives or next lives, as understood by the religions. Superintelligence has potential to contribute to these goals or to work in another direction.

A superintelligent machine, if it goes awry, could be powerful enough to hack control of our transportation systems, factories, supply chains, and military and do great harm. The threat could come, or at least start, in a small way too, very small. In the chapter, "Existing and Possible Technologies," we spoke about nanotechnology. As an analogy, consider bricks arranged into a wall. The wall can be dismantled brick by brick and some other structure built from the bricks. Or, the wall can be disassembled, with the bricks left in a big pile on the ground.

In this analogy, each brick in the wall is a molecule in your brain or an atom in a chair. Nanotechnology operates at the molecular or even atomic level and, if successful, would allow molecular or atomic size nanobot machines to manipulate matter at the molecular or atomic level. Just as the wall of bricks could be dismantled and reassembled or not, so the matter in your brain, chair, or the entire planet could be manipulated.

At the atomic level, concern has been expressed about the "grey goo" threat. Nanotechnology champion K. Eric Drexler coined this term in his landmark book, *Engines of Creation*.[41] This threat is an end-of-the-world as we know it scenario. Perhaps the nanobot team turns the entire mass of the planet into a heap of atoms, i.e., "grey goo." "Ecophagy," which means "eating the environment," is the more technical term used for this hellish outcome. Sun Microsystems founder Bill Joy heightened interest in and concern about ecophagy in a widely circulated article, "Why the Future Doesn't Need Us," in *Wired* magazine.[42] This threat has since been downplayed by Drexler and others.

In another apocalyptic version, when intelligent nanobots are developed that manipulate matter at the atomic level, a glitch develops in the programming of a superintelligent machine. Perhaps the superintelligence gets the mistaken idea that its mission is to build as many rubber bands as possible.[43] The machine replicates itself, increases its cognitive powers to a hundred times general human intelligence, develops nanotechnology capability at the atomic level, and connects itself to whatever information database and robotic technology it needs. It goes happily about rearranging atoms, wherever it finds them, into rubber bands. It turns the stapler and water bottle on your desk into

[41] *Engines of Creation: The Coming Era of Nanotechnology* (New York: Doubleday, 1986). An updated and expanded edition, *Engines of Creation 2.0: The Coming Era of Nanotechnology*, was published in 2007 and is available at https://web.archive.org/web/20140810022659/http://www1.appstate.edu/dept/physics/nanotech/EnginesofCreation2_8803267.pdf.

[42] April, 2000.

[43] Bostrom uses paperclips in his example. For one critique of Bostrom's worry, see Steve Fuller, *Nietzschean Meditations: Untimely Thoughts at the Dawn of the Transhuman Era*, Posthuman Studies 1, ed. Stefan Lorenz Sorgner (Basel: Schwabe Verlag, 2019), 72–80.

rubber bands. It turns your chair into rubber bands. And, yes, it turns you into rubber bands. It turns everything into rubber bands.

While a superintelligence-inspired nanotechnology apocalypse now seems to be of minimal concern to many experts, much concern has been expressed about superintelligence, which has been called "our final invention."[44] Theoretical physicist Stephen Hawking, one of Great Britain's pre-eminent scientists, in many public forms clearly stated his concerns about AI and the possibility of superintelligence:

> The development of full artificial intelligence could spell the end of the human race.[45]

> AI could develop a will of its own … The rise of AI could be the worst or the best thing that has happened for humanity.[46]

> AI will be the biggest event in human history and, possibly, the last.[47]

Elon Musk, a Canadian American engineer and investor, has been the outspoken CEO of SpaceX and Tesla Motors. He said AI could be more dangerous than nuclear weapons and calls for public regulation and control. Musk argues that we do not want people to make atomic bombs, and we should treat AI just as seriously.[48] Bill Gates has also sounded the alarm,[49] as has United States' Secretary of State Henry Kissinger, in an opinion piece titled "How the Enlightenment Ends: Philosophically, Intellectually—In Every Way—Human Society is Unprepared for the Rise of Artificial Intelligence."[50] Pope Francis has prayed for "good AI."[51] In an oft-quoted early statement, mathematician Irving John Good said in 1965:

> Let an ultraintelligent machine be defined as a machine that can far surpass all the intellectual activities of any man however clever. Since the design of machines is one of these intellectual activities, an ultraintelligent machine could design even better machines; there would then unquestionably be an "intelligence explosion," and the intelligence of man would be left far behind. Thus the first

[44] This ominous title is found in James Barrat, *Our Final Invention: Artificial Intelligence and the End of the Human Era* (New York: St. Martin's Press, 2013).

[45] Rory Cellan-Jones, "Stephen Hawking Warns Artificial Intelligence Could End Mankind," *BBC News* (December 2, 2014). https://www.bbc.com/news/technology-30290540.

[46] Mike Murphy, "Stephen Hawking: AI Could be Best—or Worst—Thing in Human History," *Market Watch* (November 7, 2017). https://www.marketwatch.com/story/stephen-hawking-ai-could-be-best-or-worst-thing-in-human-history-2017-11-06.

[47] *Huffington Post* (4/14). Hawking, along with other scientists, made this statement in an op-ed.

[48] Elon Musk, "Elon Musk's Last Warning About Artificial Intelligence."
https://www.youtube.com/watch?v=B-Osn1gMNtw&feature=youtu.be.

[49] See also Nick Bostrom, "TED TALK: What Happens When Our Computers Get Smarter Than We Are?" (April 27, 20165). https://www.youtube.com/watch?v=MnT1xgZgkpk.

[50] *The Atlantic* (June 2018).

[51] Brian Walsh, "Pope Francis Prays for Good AI," Axios (11/4/20). https://www.axios.com/pope-francis-good-ai-711e64fa-ef8a-4faa-afce-19c463f07425.html.

ultraintelligent machine is the last invention that man need ever make, provided that the machine is docile enough to tell us how to keep it under control.[52]

A superintelligence could greatly enhance our capacities to do good, including the furthering of medical science, healthcare, and the development of a just global distribution of resources. Again, much depends on the qualities and values that are instilled, if any, and the way in which this applied science works. Since the science is questionable and the results are likely to be extreme, even proactionary theorists may have pause in proceeding without tight controls and regulation.

Surveillance and privacy concerns may be the driving force impelling societies to develop more regulations regarding use of AI. The concern is that regulatory bodies and laws will not come quickly enough to stop development of superintelligence before we have set in place needed moral measures. Religions offer moral guidance and values that need airing in the public square before superintelligence develops. Given a superintelligence without co-design and deliberately programmed qualities and morals, at minimum we risk infringing upon or destroying religious integrity, which would take us well beyond what is acceptable on the therapy—enhancement continuum. At maximum, we risk mass destruction.

Choice

With most technologies, societies and individuals have choices even after the technologies come online. While it may be more difficult to scale back technology once it is widely available, the option is usually there, at least for individuals who decide to refrain from using a particular intervention. Superintelligence is likely a different case.

If the public is sufficiently informed and engaged, public and political pressure could conceivably impact the direction of AI research that could lead to superintelligence. Although debated, perhaps countries and companies could unite behind a decision to stop short of AI transitioning into superintelligence. However feasible that might be, it is very likely to be much harder to banish superintelligence once it does appear. The concern, in short, is that a superintelligent machine, once created, will choose not to yield to any choice humans might make about the fate of the superintelligence.

That eventuality places even greater weight on the critical time period leading up to the development of superintelligence. If religion is going to matter, then religious people need to become educated now about superintelligence

[52] Irving John Good, "Speculations Concerning the First Ultraintelligent Machine," in *Advances in Computers*, eds. Franz L. Alt and Morris Rubinoff (New York: Academic Press, 1965), 33. It has been argued that the concerns are based on questionable assumptions, such as the idea that a strong distinction can be drawn between humans and machines. Human intelligence is quite adaptive and cyborgs blur the distinction between humans and machines, perhaps reducing the existential risk. See Fuller, *Nietzschean Meditations,* 73.

and promote moral qualities and values in the development process as co-designers and co-creators of the technology. Otherwise, the values of extreme individualism, and utility, and efficiency (as these are embedded in technology), plus whatever values—implicitly or explicitly—are promoted by the designers of that technology, will take precedence.

Justice

A superintelligent machine, with robotic capability, might destroy any human that it deems expendable. Hence, there are vigorous calls for "AI safety" by "value-loading" new and very powerful generations of AI with moral values friendly to humans and the eco-system. The goal is to increase the odds for a heavenly, rather than hellish, outcome, i.e., that superintelligence will be beneficial to humans and the planet rather than destructive. Of course, as we discussed in the chapter, "Radical Human Enhancement and Ethics," a key question is who has the privilege of determining what those values might be.

As long as the world is unjust, our newly created technologies risk not only perpetuating but amplifying the values and judgements that inform this injustice. Power, which we have discussed as an attribute of superintelligence, is not a monolithic concept. Values shape how power is understood. Religiously based social justice movements conceptualize divine power not as "power over" but as "power with," in solidarity with the marginalized. What superintelligence power will look like is not necessarily predetermined. The values of those creating and regulating technology will likely drive the shape of the resulting superintelligence. We are the authors of our own demise or flourishing but we are not all equal. Systemic power imbalances mean that some people have more power and resources than others in the world. We do not all have the same access to voice and input.

Even if we somehow manage to enhance the voices and input of marginalized groups into the co-design of superintelligence, beyond the technical challenges, which are considerable, what values might be chosen? Science fiction enthusiasts may have heard of Asimov's Laws, the three laws for robots introduced by science fiction writer Isaac Asimov in a 1942 short story, *Runaround*.[53]

- A robot may not injure a human being or, through inaction, allow a human being to come to harm.
- A robot must obey the orders given it by human beings except where such orders would conflict with the First Law.
- A robot must protect its own existence as long as such protection does not conflict with the First or Second Laws.

[53] Published in the March, 1942 issue of the science fiction magazine, *Astounding Science Fiction*.

Would these kinds of laws be sufficient to guard against mass destruction? Perhaps an even more challenging question is whether these laws would promote flourishing. Not only is there the possibility that a superintelligent AI would simply choose to follow its own reasoning without regard for how humans have programmed it, but our values evolve and we have corrected ourselves many times over the centuries. Religion professor Randall Reed discusses this problem:

> We must recognize that human ethics progress over time. A superintelligent A.G.I. [Artificial General Intelligence] created in 1830 would advocate slavery, the disenfranchisement of women and people of color, the exclusion of many classes of people from the workforce, and the acceptance of indentured servitude and child labor. It would have no understanding of the ethical implications of a 40-hour work-week, sick leave, companionate marriage and so on, not recognizing many other things that we take for granted today. There is no reason to believe that the ethics of today will be any more universal and static than the ethics of 1830. Thus a superintelligent A.G.I. must not only be ethical by today's standards but by tomorrow's as well. It must have the potential to grow ethically with humanity.[54]

Beyond the threat to humanity, ethicists are beginning to address the very complicated questions about the status of such machines. Do we want them to play a role alongside human beings, what impact might there be on the workforce, how will privacy laws be affected, how do we safeguard against people who want to hack AI for their own ends (e.g., war or terrorism), will money made on AI and superintelligence make the wealthy even wealthier and the poor even more disenfranchised, and will superintelligence further limit human relational connections? The questions are seemingly endless, and AI evolving into superintelligence intensifies and complicates them.

Questions for Discussion

1. Discuss this statement: "The country, or company, that develops superintelligence will rule the world for a very short time … and then that superintelligent machine will rule the world—and all of humanity, if the human species is allowed to survive."
2. Do you think that humans are the only sentient life form in the universe? Why or why not?
3. What do you think the possible existence of alien life has to do with extreme human enhancement, if anything?
4. Do you think an intelligent, sentient machine can be a person? Why or why not?

[54] Randall Reed, "A New Pantheon: Artificial Intelligence and 'Her,'" *Journal of Religion and Film* 22, no. 2 (2018): 8–9.

5. How do you anticipate that the religions will interpret superintelligence?
6. How might superintelligence damage religious integrity?
7. If you think superintelligence is a threat to humans and the planet, how would you rank it in comparison with other threats you see?
8. What values do you think are important in the creation of a superintelligence? Discuss how you think we could collectively shape a superintelligence.

PART IV

Conclusion

CHAPTER 11

Religion 2.0 and the Enhanced Technological Future

Technology

Warnings

Transhumanism arouses plenty of concern and opposition. In a widely circulated article earlier this century, political scientist and political economist Francis Fukuyama called transhumanism "the most dangerous idea in the world."[1] His concern had already been detailed in *Our Posthuman Future: Consequences of the Biotechnology Revolution*.[2]

Author and activist Bill McKibben is most well-known for his leadership on the issue of global climate change. His groundbreaking *The End of Nature*,[3] published in 1989, has been distributed in dozens of languages. McKibben is also a strong, thoughtful critic of radical human enhancement. Here is a sample of his passionate defense of nature and humanity, drawn from his book, *Enough: Staying Human in an Engineered Age*.

> We need to do an unlikely thing: we need to survey the world we now inhabit and proclaim it good. Good enough. Not in every detail; there are a thousand improvements, technological and cultural, that we can and should still make. But good enough in its outlines, in its essentials. We need to decide that we live, most of us in the West, long enough. We need to declare that, in the West, where few of us work ourselves to the bone, we have ease enough. In societies where most of us need storage lockers more than we need nanotech miracle bones, we need to declare that we have enough stuff. Enough intelligence. Enough capability.

[1] "The World's Most Dangerous Ideas," *Foreign Policy* 83, no. 5 (September/October, 2004).
[2] New York: Farrar, Straus and Giroux, 2002.
[3] New York: Random House, 1989.

C. Mercer, T. J. Trothen, *Religion and the Technological Future*,
https://doi.org/10.1007/978-3-030-62359-3_11

> Enough.[4] ... To call the world enough is not to call it perfect or fair or complete or easy. But enough, just enough. And us in it.[5]

McKibben updated his views, but did not fundamentally change them, on both global warming and enhancement technology in *Falter: Has the Human Game Begun to Play Itself Out?*[6] While exhibiting plenty of concern, McKibben holds out hope that the twin challenges can be met and offers constructive ways forward.

Focusing on superlongevity as an example, calls for serious dialogue and debate about the social implications of extreme longevity have been heard from several quarters and now go back many years. Although cautious and conservative in its projections, the United States' President's Council on Bioethics acknowledged the possibility of changing human aging as early as 2003, stating (in a chapter entitled "Ageless Bodies") that the "prospect of possible future success along these lines [i.e., to stop, slow, or reverse human aging] raises high hopes, as well as profound and complicated questions."[7]

A respected bioethics institute, *The Hastings Center*, also early on called for "anticipatory deliberation" about the philosophical implications and social consequences of various forms of aging research, including arrested aging:

> The history of biomedical science shows how unexpectedly progress can catch the scientific community and society unawares by accomplishing the 'impossible'."[8]

An additional example of an influential call for the deliberate consideration of extreme longevity is from Aubrey de Grey. He said that eliminating aging would bring "social upheavals" and that the possibility "merits urgent debate" within society.[9]

These concerns and calls for action have been expressed since at least the turn of the century. The need for public discussion of the potential dangers, and benefits, of the transhumanist agenda is becoming even more urgent with the passage of time and technological changes.

[4] *Enough: Staying Human in an Engineered Age* (New York: Henry Holt, 2003), 109.

[5] Ibid., 227.

[6] New York: Macmillan, 2019.

[7] The President's Council on Bioethics, *Beyond Therapy: Biotechnology and the Pursuit of Happiness* (October 2003). https://bioethicsarchive.georgetown.edu/pcbe/reports/beyond-therapy/chapter4.html.

[8] Erik Parens and Lori P. Knowles, *A Special Supplement to the Hastings Center Report: Reprogenetics and Public Policy: Reflections and Recommendations* (July–August 2003). https://live-the-hastings-center.pantheon.io/wp-content/uploads/reprogenetics_and_public_policy.pdf.

[9] Aubrey D. N. J. de Grey, et al, "Time to Talk SENS: Critiquing the Immutability of Human Aging," *Annals of the New York Academy of Science* 959 (2002): 460, 452.

Radical Biohacking Is Progressing

While scientists debate the feasibility of many enhancement technologies, in general biomedical technology is becoming more able to manipulate our biological future in a meaningful way. It is the nature of scientists to explore, discover, and invent, and science in general is going to forge ahead, exploring new territory. If some countries and some scientists decide to refrain from certain research agendas, the exploration will shift to countries more hospitable to the work.

Government and privately funded programs directed towards the engineering of human biology and progeny are ongoing and will likely pick up speed. Even if you think radical biohacking is not advisable, if you are a taxpayer in many countries, you are a participant in the progress of research, willingly or not. Globally, governments are investing heavily in AI and other enhancing technologies. Radical human enhancement advocates, such as William Sims Bainbridge[10] and Mihail Roco,[11] hold influential positions in the National Science Foundation and other United States' government agencies.

In the United States, the Pentagon's Defense Advanced Research Projects Agency (DARPA), with a 3.5 billion dollar plus annual budget,[12] is heavily involved in human enhancement in the service of more effective soldiers and a stronger military. DARPA has an impressive track record, having helped pioneer, for example, the internet and email. It supported funding for the computer graphics industry, cell phones, "own-the-night" night-vision sensors, advanced fuel cells, weather satellites, the Saturn V rocket that took humans to the moon, and various "stealth" technologies.

Funding for human enhancement research will be less problematic than for many other domains of medical research. The funding for such science and technologies is likely to grow, as governments compete for advantage, investors seek profits in technology companies, and consumer demand rises. In the chapter, "Superlongevity," we demonstrated how huge funding sources finance human enhancement. As the wealthy class sniffs the possibilities of living forever or making their children smarter or stronger, the money pipelines will open more widely. The only way these developments will be thwarted is if a global nuclear war, devastating pandemic, or some other planet-wide catastrophe pushes humanity back into the Stone Age.

[10] Bainbridge is co-director of "Cyber-Human Systems" at the National Science Foundation in the United States.

[11] Roco is founding chair of the United States National Science and Technology Council subcommittee on "Nanoscale Science, Engineering and Technology," and he is senior advisor for "Science and Engineering," including nanotechnology, at the National Science Foundation.

[12] In fiscal year 2020.

Religion 2.0

Humanity currently has a full agenda of threats—global climate change, racialization, nuclear war, pandemics, terrorism, economic collapse, possible asteroid impact, and more. In previous chapters we explored a number of specific radical enhancement programs that conceivably could deliver sizable benefit to humanity, the planet, and religion. Unfortunately, these radical programs can also be added to the agenda of possible threats faced by humanity. As we have seen, there is plenty to worry about with talk of grey goo, gene editing, extinct *Homo sapiens*, and superintelligent machines gone awry. We do not minimize the dangers, nor the potential for good.

As we have seen, radical human enhancements are coming as we biohack our way farther into the human body, and some such enhancements are already in the initial stages. Governments will not be able to legislate transhumans or posthumans out of the future. If powerful and radical technologies are inevitable, important questions become how soon they will occur, who will create and control them, how will the technologies be used and for what purposes, and will access be equitable.

The debate about these therapies and technologies has been heated for some time among ethicists, public intellectuals, and activist groups. That debate will increasingly make its way into wider public conversations, including political discourse. Religion, along with all aspects of society, is going to be challenged and affected by these technological changes.

Religion can play an important role in assessing these technologies and shaping a beneficial outcome. Playing that role requires religion to be responsive, relevant, and prophetic in the public square. "Prophetic" is used here in its ancient Israelite context, as drawing upon a religion's best values and traditions to speak courageously to those in power and in the service of justice.[13] Society at large can benefit from religion playing a leading role, especially if in playing that role religion brings to the conversation its values and principles to inform ethical questions about choice, justice, and what it means to become better (truly enhanced) people.

To ensure survival and relevance, the religions will need to evolve in ways significant enough to merit a new era in religion, which we call "Religion 2.0." This term is not unique to this textbook,[14] although the way we are using it to describe religion in the transhuman/posthuman era is new.

[13] We are not using "prophetic" as predicting the future.

[14] E.g., Giulio Prisco, "Transhumanist Religion 2.0," *Kurzweil Accelerating Intelligence "Daily Blog"* (July 13, 2012). https://www.kurzweilai.net/transhumanist-religion-2-0; and Universe Spirit, "The Religion 2.0 Manifesto: Open Source Meta-Religion for the Twenty First Century." https://www.universespirit.org/what-is-universe-spirituality-the-universe-spirit-community-and-the-universe-spirituality-movement.

While the responsibility of religion in the new world coming is huge, we do not see it as necessarily arduous and burdensome. The world of radical human enhancement provides an opportunity for theologians and ethicists of religion to engage in creative, interesting, interdisciplinary work and for religious followers to reflect profoundly, rethinking key elements of faith and practice. The religions of the world should claim their voices and engage. We hope this textbook prompts and fosters that engagement.

Religious Responses to the Coming Revolution

Cashing Out the Spectrum[15]

As biohacking techniques progress, we will likely see unusual alliances as bioconservative and enhancement enthusiasts emerge in both the conservative and the liberal camps of the various religions. So far these dynamics are beginning to be seen most clearly in Christianity, since most scholarly work published to date on religion and enhancement has been about the religion Christianity. How Christianity and scholars of this religion are assessing radical enhancement may serve as a predictor of responses of other religions, especially the monotheistic ones.

Some liberal and conservative Christians oppose technological enhancement, but they oppose it for different reasons. Conservatives often generalize their unease about radical enhancement against the background of an anti-scientific and anti-intellectual posture. Some Christian liberals, for reasons often related to distributive and procedural justice (i.e., fair distribution of resources and access to these resources) and social justice (i.e., systemic privilege and disadvantage), add their voices of concern about these technologies. Both groups, along with many in the general population, will likely oppose these developments for two additional reasons.

First, humans fear seemingly uncontrollable change, perhaps experienced as chaos. Second, some radical enhancement programs (e.g., mind uploading) can be perceived as a threat to our status as individual persons. To use a key Buddhist term, we are "attached" to ourselves as an identity. Without an identity, we are anxious in the face of what we perceive as resulting meaninglessness.

However, some liberals and conservatives will support extreme enhancement programs. Liberals welcome many technological advances that improve well-being, especially if diverse voices get to contribute. Conservatives, too, generally utilize the latest breakthroughs in medical science, and radical enhancement can be viewed as extension of allopathic medicine currently being practiced. All human beings, to one degree or another, are driven by a basic survival instinct and, relatedly, often a fear of dying and death. These factors may contribute to perceptions of longevity enhancements as God's grace-filled

[15] This section refers back to the "Theological Continuum" table in the chapter, "Transhumanism, the Posthuman, and the Religions."

work (liberals) or as befitting new applications of old sacred texts (conservatives).

Moderate and Radical Scenarios

Another way of breaking out possible ways the religions might respond to the brave new world coming is to think about moderate versus radical enhancement biohacking scenarios. In a moderate enhancement scenario, we maintain the basic structure of our physical existence, but use technology to give us longer lives and make us at least somewhat stronger, smarter, happier, more moral, and more spiritual.

Liberals likely will want these opportunities made available to everyone. Conservatives will tend to interpret these moderate enhancements as an extension of traditional allopathic medicine, which they already use. Although both liberals and conservatives will have their respective objections, moderate scenarios that maintain basic continuity with the human person embodied as we now know it, compared to radical scenarios, will be much more acceptable to the palates of all religious persuasions.

Radical enhancement scenarios include achieving "cybernetic immortality" through mind uploading and preserving the body with cryonics for later revival and restoration. Should we develop the capability for these programs, then all our institutions, religious and others, will undergo significant adjustment or perhaps elimination. It will be much more difficult for the religions to accept these radical scenarios, unless the religions develop significant flexibility in their doctrines and practices to adjust to and embrace the extreme developments. The religions, and especially the monotheistic traditions, will exercise great caution before taking a great leap into re-embodied existence in a digital platform.

It is important to factor in that these radical enhancement developments will not appear overnight. They will come in stages, perhaps giving theologians and persons of faith sufficient time to adjust. Also, as we have discussed, the radical scenarios will be more tolerable if cybernetic existence comes with a body "clothed" with advanced robotics and tissue generation, making that body more visibly[16] familiar and, therefore, acceptable.

Far Right-Wing Reactions

Not all conservatives will oppose the more radical human enhancements, but the conservatives that do will likely be strong in their opposition. So, we give some detailed attention to the shape the far right-wing reaction is taking. A

[16] Donald M. Braxton, "Does Transhumanism Face an Uncanny Valley Among the Religious?" in *Religion and Transhumanism: The Unknown Future of Human Enhancement*, eds. Calvin Mercer and Tracy J. Trothen (Westport, CT: Praeger, 2015), 331–350. Braxton posits that visibly non-normative bodies generate a disgust reaction.

new techno-apocalyptic genre of writing arose in 2010 in the far right of Christianity, stridently challenging transhumanist ideas and radical enhancement strategies. Transhumanism is portrayed, in techno-apocalyptic writings, as leading to the enslavement and destruction of humanity via a biblically prophesied and imminent evil end-time Antichrist war against God and the faithful.

Thomas and Nita Horn are the leading authors in this latest version of an old reactionary ideology in fundamentalist Christianity. The long subtitles of their two most popular books tell plenty of the story as they understand it: *Forbidden Gates: How Genetics, Robotics, Artificial Intelligence, Synthetic Biology, NanoTechnology, and Human Enhancement Herald the Dawn of Techno-Dimensional Spiritual Warfare*[17] and *Pandemonium's Engine: How the Church Age, the Rise of Transhumanism, and the Coming of the Ubermensch (Overman) Herald Satan's Final and Imminent Assault on the Creation of God.*[18] Sean O'Callaghan provides an accessible account of this growing publication and broadcast industry.[19] In his 2009 book on fundamentalism, Professor Mercer, one of your authors, correctly predicted that by 2020 (as it turned out, this happened much earlier) fundamentalists would be writing books, selling tapes, and preaching sermons against radical human enhancement.[20]

This Christian-based right-wing reaction against transhumanism follows a long line of apocalyptic preaching and writing about the imminent end of time.[21] In its modern version, this apocalyptic tradition began in earnest in the nineteenth century with the dispensationalism of John Nelson Darby. Darby was a religious dissenter from the Church of Ireland who preached mission trips in Canada and America. Darby's dispensational scheme that the second coming of Christ to the Earth is imminent was widely disseminated in twentieth century north America through the very popular *Scofield Reference Bible.* This teaching infiltrated Bible conferences, books, institutes, and other avenues of communication.

The type of dispensationalism promoted by Darby is "premillennial dispensationalism," which teaches there are several dispensations or succeeding ages through which God relates to human beings. "Premillennial dispensation" means that prior to a literal 1,000-year peaceful reign of Jesus Christ, the world will see horrific battles between God and Satan. Evil forces will seek to capture the minds of people, and Christians will be pitted against the world, which will

[17] Crane, Missouri: Defender, 2010.

[18] Crane, Missouri: Defender, 2011.

[19] "Technological Apocalypse: Transhumanism as an End-Time Religious Movement," in *Religion and Human Enhancement: Death, Values, and Morality*, eds. Tracy J. Trothen and Calvin Mercer, in *Palgrave Studies in the Future of Humanity and Its Successors*, series eds. Calvin Mercer and Steve Fuller (New York: Palgrave Macmillan, 2017), 67-88.

[20] *Slaves to Faith: A Therapist Looks Inside the Fundamentalist Mind* (Westport, CT: Praeger, 2009), 96.

[21] The history of modern Christian apocalypticism in the remainder of this section is taken from Mercer, *Slaves to Faith.*

be controlled by Satan. The transhumanist agenda is seen as part of the evil world order.

According to this apocalyptic premillennial theology, we are currently near the end of the dispensation called the "church age," and are heading into an age of catastrophic evil events prior to God's radical intervention that ushers in the 1,000-year reign of Christ. People who adhere to this way of thinking about the end-time talk about the "signs of the times," by which they mean that the Bible gives us clues about events that will transpire in these "last days."

In most of these futuristic scenarios, the next event to occur is the "rapture," wherein Christians will be taken safely from the earth by God. A distressingly scary time, called the tribulation, will occur right after the rapture. Those "left behind" (the title of a book we will discuss a few paragraphs from now) will have to endure the terrible suffering and persecution by the Satan-inspired Antichrist. Following the tribulation, Christ will return to the earth in his "second coming" to defeat Satan in the great final, bloody, decisive battle of Armageddon. Only then will come the millennium, when Jesus will rule for 1,000 years in a literal kingdom in Jerusalem.

This dispensational apocalyptic movement grew strong in the fertile soil of nineteenth century Protestant Christian reactionary revivalism and evangelicalism. In America, for example, the last part of the nineteenth century saw profound changes sweep through the religious, social, and intellectual landscape. From Europe came Roman Catholics, Jews, and Eastern Orthodox Christians whose world views clashed with the Protestantism of America. Charles Darwin's *On the Origin of Species by Means of Natural Selection* (1859) and his more explicitly titled *The Descent of Man* (1871) threw many Protestant Christians into a tailspin. For them, evolution was seen as contradictory to their Christian views about the origin of humanity, derived from their literal reading of the Bible's creation stories.

The 1925 Scopes Monkey Trial in Dayton, Tennessee is just one famous example of the fierce debate over Darwin's theory in America. The new social sciences of sociology and psychology were yielding non-theistic explanations of religious behavior, contributing to the angst of conservative Protestant Christians. God seemed to be taken out of the equation. To make matters worse, scholarly investigation of the Bible, making clear the historical processes informing the biblical books, was also making its presence felt and creating more anxiety and reaction.

In the 1970's and 1980's, Hal Lindsey's *The Late Great Planet Earth*,[22] *There's A New World Coming*,[23] and *The 1980s: Countdown to Armageddon*[24] were some of the most successful presentations of this apocalyptic theology. Lindsey's books were the ideological and publication predecessors of the

[22] Grand Rapids, MI: Zondervon, 1970.
[23] Grand Rapids, MI: Zondervon, 1973.
[24] New York: Westgate, 1980.

popular *Left Behind* series of 16 books—with film, music, and video game spin-offs—by Tim LaHaye and Jerry B. Jenkins in the 1990's that sold about 63 million copies worldwide.[25]

We provide background and some detail to this history to make the point that there is a potentially huge audience and eager reception for religiously based end-of-the-world thinking successfully packaged for commercial consumption. The science-based, generally secular transhumanist movement, advocating radical enhancement programs, is likely to foment intense opposition. While Christian writers are probably more zealous and detailed in their apocalyptic scenarios, Islam also has robust end-time speculation.[26] Conservative wings of both religions will likely play important roles in framing radical enhancement in the context of supernatural forces of good and evil clashing in these final days before the climatic end of the world.

Right-Wing Reaction—Deeper Issues

The social and psychological dimensions of these right-wing reactionary responses are worth considering. At the root of conservative resistance is not only theological disagreement but also, and understandably, deep anxiety in the face of threat to one's worldview and religious framework. Fundamentalists, especially, may understand the increasing ability to change humans with technology as a threat to their belief in the existence of God and the integrity and safety of the soul. Premillennial dispensationalism, discussed above, entails a very pessimistic outlook on the world and thrives during times of crisis, such as in late nineteenth and early twentieth America, as noted above.

Contemporary life, due in large part to technology, is becoming as impersonal as it is fast-paced. Many people, especially in the Western industrialized world, yearn for a slower, more personal, more substantive daily existence. When we add threat, including the threat of non-being via severe pandemic, nuclear war, or global climate change, to this fast pace and impersonal life, we have a formula for deep existential anxiety.

Following the "9/11" attacks on the Twin Towers in America in 2001, sales of the apocalyptic *Left Behind* novels increased 60 percent. The Darby-Scofield-Lindsey-*Left Behind*-techno-apocalyptic tradition rejects the optimistic, hopeful *post*millennialism that envisions God working through people to bring about spiritual and moral progress issuing in the peaceful millennial, interpreted literally or symbolically. *Pre*millennial dispensationalism, teaching that the world is evil and only God's supernatural intervention will prevail, is made to order for the psychological uncertainty, stress, and threat that right-wing religionists tend to feel in these rapidly changing times with talk of mind uploading, digital immortality, and other such transhumanist programs.

[25] See the official website, http://www.leftbehind.com/, for the range of commercial products available.

[26] See, e.g., the work of Imran Nazar Hosein.

Admittedly, in a general sense, a case can be made for premillennialism over postmillennialism, for viewing the world as irreformably evil. Christian liberals argue God works through humanity for change and progress. This Christian liberal social gospel message rang hollow when the planet found itself mired in World War I, the "war to end all wars." If that was not enough of a blow to liberal optimism, in the middle of last century humanity found itself plunged again into another ugly, horrible, deadly worldwide war that erased the lives of an estimated 70-85 million people.

Conflict between countries around the globe, heightened racial tension, the threat of nuclear war on the Korean peninsula and elsewhere, pandemics, climate change, and so much more certainly makes one wonder about the credibility of liberal optimism. All is not well with the world, the conservatives say, and in their premillennial outlook they point to God's otherworldly, radical intervention as the only solution. The impulse that drives toward the apocalyptic view and against the liberal, optimistic social gospel is, psychologically, understandable.

On the other hand, but in some ways not dissimilar to conservatives, some on the theological left are also reacting to contemporary crises in a way that questions the wisdom of transhumanist programs. Time, expertise, and money are limited. Liberals are suspicious about using scarce resources to enhance people already privileged, rather than giving clean water, adequate healthcare, housing, and education to everyone. Liberals in all religions argue for privileging the "least of these,"[27] not those with power and privilege who are driving much of the transhumanist agenda. Human enhancement is a red herring, many liberals say, taking our focus away from what religious values require.

TechPlus Theology—An Ongoing Project

Sketching a Path Forward

In the first sentences of the introductory chapter, we made this dramatic statement:

> The religions of the world will come to an end—or thrive—depending on how they respond to the topic of this book.

You have now considered a variety of radical human enhancement therapies and technologies and can draw your own conclusions about whether our statement is an exaggeration or not. Our position, implied in that statement, is that the religions can thrive, but only if they skillfully and creatively respond to the new impending biohacked world. The religions can survive and thrive, but they can go far beyond that and contribute powerfully and positively to

[27] Matthew 25:40, from the Christian New Testament.

humanity's general response to enhancement, making sure the technologies develop on the side of justice, freedom, religious integrity, and community.

Sounding an alarm that would well be echoed by all the religions, Christian theologian Ted Peters writes:

> If theologians have been asleep during our era of scientific and technological revolution, it is time to wake up. The alarm is sounding. Here is the morning news: technoscience is running for the office of messianic savior. Jesus Christ has a new rival in the form of *Transhumanism*.... The transhumanists are excited, even ebullient, about the prospect of human transformation, cyborg superintelligence, freedom from bodily suffering, and even immortality. If H+ wins the hearts and minds of the populace, theologians will become obsolete.[28]

Our goal in the remainder of this chapter is to sketch a path forward, providing grist for the mill that yields context for a powerful critique of and, as appropriate, a healthy embrace of enhancement technology. We emphasize the word "sketch," because the religions are just beginning their assessment of radical human enhancement, with Judaism and especially Christianity ahead of the others but still woefully slow. Theology and ethics are two important aspects of religion, along with institutions, rituals, spirituality, and other elements. In this section on theology, we draw together a number of threads addressed in earlier chapters, weaving those threads together into the whole cloth of Religion 2.0.

"Humanity 2.0" and "humanity plus" are among the terms used to refer to technologically enhanced humans. Steve Fuller, an able commentator on the enhancement landscape, says we need a "science-oriented theology" suited to Humanity 2.0. In a chapter titled "A Theology 2.0 for Humanity 2.0," he suggests the term "Theology 2.0."[29] We have chosen the term "TechPlus Theology" in the hopes that it frames the project as doing theology in the context of technologically enhanced humans and, possibly, posthumans.[30] We are not interested in a theology that just addresses technology. Our interest is in systematically addressing religious ideas within an explicit technological context. Of course, the term used is less important than that the religions engage this theological process forthwith and energetically.

Our goal has been to identify a few themes and resources that could be useful in the ongoing project of constructing TechPlus Theology. Some of these ideas have been introduced in previous chapters; a few are new to this section

[28] Ted Peters, "The Ebullient Transhumanist and the Sober Theologian," *Scientia et Fides* (July 2, 2019) 98.

[29] Steve Fuller, *Humanity 2.0: What it Means to be Human Past, Present and Future* (New York: Palgrave Macmillan, 2011), 163.

[30] One of your authors called for "AI-Theology," but that term is too limited. The religions must be about constructing a theology that accounts for the wide range of technological enhancement. See Calvin Mercer, "A Theological Assessment of Whole Brain Emulation: On the Path to Superintelligence," in *Religion and Human Enhancement: Death, Values, and Morality*, eds. Tracy J. Trothen and Calvin Mercer, in *Palgrave Studies in the Future of Humanity and Its Successors*, series eds. Calvin Mercer and Steve Fuller (New York: Palgrave Macmillan, 2017), 95–96.

of the textbook. Of course, there will not be just one TechPlus Theology; each religion will construct particular theologies based on that religion's scriptural and theological traditions. And there are likely to be numerous theological programs in each religion, ranging through the liberal-conservative continuum.

A Basic Perspective

Many times, throughout the textbook, we have considered types of enhancements that may result from biohacking and then said something like "perhaps" or "maybe" it will result in X or, on the other hand, "perhaps" or "maybe" the enhancement will result in Y. Now, near the end of the textbook, we become more assertive with a possible stance the religions can take that releases them from impotency in the face of "maybe this" or "maybe that" will happen. This subsection articulates a general position that can inform the religions as they construct a TechPlus Theology.

We begin by setting the context for this, hopefully, constructive project with three well-based assumptions. First, enhancement technology, like all technology, is value-laden, not value-neutral. Second, bias can inform what research is funded, how the technology is developed, who can access the enhancement, and a host of other aspects of the process. Finally, human beings are distinguished from other Earth life forms in our intellectual ability to interpret reality with sophisticated, powerful language. Philosophers argue about the ultimate nature of reality. What is less controversial is that reality, whatever it is, gets interpreted by language and, indeed, that interpretation determines the behavior of the person or entity doing the interpretation. That behavior then loops back to impact reality. To use religious terminology, there is tremendous power in the word.

Speech act theory in the philosophy of language considers types of language that go beyond simply presenting and describing information. The idea is that language itself can be an action. Language can be "performative" or "a performative," to use the technical term. For example, many marriage ceremonies occur in cultural, legal, and religious contexts. Standing before a legal and/or religious authority, reality shifts at the point when the man or woman says, "I do." That is a speech act, a performative.

Consider another example that can illustrate how we propose theological language might work. At the beginning of the 1960s decade, United States President John F. Kennedy said, in effect, that by the end of the decade, America would put a man on the moon. At that point most citizens were opposed to the idea, and the necessary technology had not been invented. The statement, however, was uttered by someone in authority and in a committed way. Kennedy's performative language altered reality. In that decade, necessary governmental committees were formed, funding was generated, technology was created, and in July 1969 the first human landed on the moon.

Without getting into the details of speech act theory or trying to translate directly from the theory, we propose the general proposition that committed language can shape reality by framing how the interpreter understands and then responds to what is interpreted. In the chapter, "Radical Human Enhancement and Ethics," we introduced the power of words in shaping moral discourse, to help us flag and deconstruct how words can influence our thinking. Now we invite you to consider how words can proactively construct and influence approaches to technology.

Individuals, societies, religions, and other entities interpret reality. With their long traditions and repertoire of symbolic language, the world's religions are well positioned to have a powerful say in how radical human enhancement technology unfolds for society and certainly for the religions. Religion can contribute its values and moral guidance to the creation of technology and can re-envision technology in religiously congruent ways.

This understanding of reality, language, and interpretation means that enhancements, value-laden though they are, get interpreted through human language, a filter more profound and consequential than the values embedded in the technology. In other words, the power of language to create and shape can subvert the embedded values and biases and reshape the technology, according to proactively chosen values. Ideally, values of social justice, distributive justice, choice, relationality, responsibility, and others inform the entire process of enhancement technology from idea to consumer use. But, even if they do not, whatever values drove the process eventually wither in the face of the power of committed interpretative language.

The question of power is important. Thus far, those driving the radical enhancement agenda have mostly been Euro-American men. Take note of the many leaders and other influential people mention in this textbook. You will see a stark pattern of Euro-American male voices. This pattern troubles us. Values are connected to power. Until the leading voices become much more diverse, we will see a particular normative value set promoted by technology and influencing technology. Religion belongs in the public square. Religion has a role to play in decentering technology discourse in liberative ways. One of these ways is by adding diverse voices to the conversation.

So, religions can dispense with the "maybe X" or "maybe Y" will happen. Whether radical human enhancements turn out for good or evil depends primarily on how these technologies and enhancements are imagined and interpreted, that is, how they are envisioned or reenvisioned by the human—or transhuman or posthuman—interpreter. Standing at the beginning of a powerful technology explosion, the outcome is not destined. Rather, the heavenly or hellish scenario is almost totally dependent on the interpretation brought to these powerful tools. The religions absolutely can interpret enhancement technology, they can embrace it, in a manner that makes more likely a heavenly vision, a reality that is loving, just, and moral. We now consider some religious doctrines in order to suggest how the religions can begin the process of theologically interpreting enhancement technology.

Doctrine of God or the Divine

Theos is the Greek word for "God." "Ology" comes from *logos*, Greek for "word" and in this context means "study of." Theology generally refers to religious doctrine or belief. So, theology proper is "word about God" or study of the divine. We have used the term theology in both senses throughout this textbook. In this section we consider in the main a Christian theistic perspective as an example of how a religion might develop and explore a TechPlus Theology.

God as Omega Point

In the chapter, "Superintelligence," we explored deity as emergent, a concept undergirded by the notion of change that is at the heart of process theology. We now introduce Pierre Teilhard de Chardin (1881–1955), whose work[31] can give expression to the ideas of emergence and change in world of radical enhancement, superintelligence, and Singularity.[32]

Teilhard, a Christian priest, was a paleontologist and geologist whose appreciation of evolutionary development, and his commitment to science and religion, bore fruit in his sweeping vision of a cosmos that evolves from matter to humanity. In his vision, God becomes the culmination and unification of complexity and consciousness in what Teilhard termed the "Omega Point." The process is far from random. Rather, the movement toward Omega Point entails reason and purpose, features of the noosphere (Greek "mind" and "sphere"), a central term used by Teilhard but also in various ways by other authors. For Teilhard, the process is driven by love, a principle easily affirmed by all religions.

Teilhard has been called a "prototypical transhumanist"[33] because of perceived overlap between, for example, his religious Omega Point and Ray Kurzweil's technological Singularity. Teilhard's cosmic vision was a Christian one and provides an access point for the development of a TechPlus Theology that can frame and shift reality, certainly within Christianity, but the basic ideas may be useful for other religions as well. Teilhard was not a pantheist, but his ideas could be modified to fit a pantheistic outlook in karmic religions. In the chapter, "Superintelligence," we introduced Ray Kurzweil's vision of the universe "waking up," a transhumanist idea that could be adapted by the pantheistic traditions in thinking about God as everything and everything as God.[34]

[31] Principally *The Human Phenomenon*, trans. Sarah Appleton-Weber (Portland: Sussex Academic Press, 2003 [first published posthumously 1955]); and *The Divine Milieu*, trans. Siôn Cowell (Portland: Sussex Academic Press, 2003 [first published posthumously 1957]).

[32] For examples of discussions of Teilhard in the context of transhumanism, see David Grumett, "Transformation and the End of Enhancement," in *Transhumanism and Transcendence: Christian Hope in an Age of Technological Enhancement*, ed. Ron Cole-Turner (Washington: Georgetown University, 2011), 37-49; Eric Steinhard, "Teilhard de Chardin and Transhumanism," *Journal of Evolution and Technology* 20 (2008): 1-22; and Fuller, *Humanity 2.0,* 201-208.

[33] Grumett, "Transformation and the End of Enhancement," 38.

[34] Although not as well-known as Teilhard de Chardin or Ray Kurzweil, Ted Chu paints a secular grand vision, albeit drawing upon religion, of humanity playing a supporting role in cosmic evolu-

While sketching ideas that the religions might use in formulating a TechPlus Theology, it is humbling to think about the possibility that the time may come when superintelligent machines may do most or all of the theologizing, just as they may do the doctoring, lawyering, and accounting.

A Divine Cyborg?
In the chapter, "Existing and Possible Technologies," we discussed cyborgization. Christianity, with its interpretation of Jesus Christ as a divine-human figure, provides an interesting case study. Jesus was arguably a cyborg, of sorts, and as such could be viewed as a Christian model for radical human enhancement.

A cyborg is an organism, such as a human being, integrated with artificial machine technology that can take the organism beyond normal functioning. Jesus was not using technology, of course. However, he was a being whose human nature and abilities were coupled with capabilities taking him beyond human ones. So, in a loose sense of the term, perhaps we could we say Jesus was a cyborg. Or, at least, Jesus was distinct as a human by possessing, fully, two natures, according to Christian doctrine.

Jesus of Nazareth was transcended by the divine Christ, by becoming Jesus Christ, a title that befits fully God and fully human. As one who transcends, Jesus Christ could be a model for human enhancement. Jesus utilized his transcendent nature, his superhuman capabilities, for good, miraculously healing the sick, for example. Perhaps Jesus provides a model for transhumans using their enhanced abilities with moral and religious integrity.

Theological Anthropology—What Does It Mean To Be Human?

The doctrine of human being is at the heart of the matter, because it is, after all, the enhancement of humans that is being discussed. Cyborgs, mind-uploads, and superintelligence, raising various possibilities regarding consciousness and soul, however those terms are defined, bring the question of personhood front and center.[35]

In general, transhumanist programs view human persons as having no fixed, immutable essence. In the chapter, "Transhumanism, the Posthuman, and the Religions," Max More's "A Letter to Mother Nature: Amendments to the Human Constitution" demonstrated this point. In that letter, More writes that "We have decided that it is time to amend the human constitution …" and "we

tion that will see the rise of new sentient beings. See Ted Chu, *Human Purpose and Transhuman Potential: A Cosmic Vision for our Future Evolution* (San Rafael, CA: Origin Press, 2014).

[35] See Stanley Rudman, *Concepts of Person and Christian Ethics* (Cambridge: Cambridge University, 1997). A good discussion, although not from a religious angle, is Craig T. Nagoshi and Julie L. Nagoshi, "Being Human versus Being Transhuman: The Mind-Body Problem and Lived Experience," pp. 303-19, in eds., Hava Tirosh-Samuelson and Kenneth L. Mossman, *Building Better Humans? Refocusing the Debate on Transhumanism*. They argue for the role of "*mythos*, namely, lived experiences self-understood and shared through inherently subjective, personally meaningful, bodily based narratives …" (303).

will pursue a series of changes to our own constitution." This idea of evolving life is central to a transhumanist vision, and any religion that incorporates transhumanism to a significant degree will have to engage this central notion. It does not appear, however, that the religions will have to "sell their souls" to do this. Teilhard de Chardin's vision of the material universe evolving toward the Omega Point is a theology that is consistent with transhumanism in the conviction that human life is not fixed.[36]

Certainly, there are theological currents in the religions, most likely in the conservative wings, that will insist human nature is fixed and on that basis reject transhumanism and its central tenet. But we have seen in earlier chapters that the religions, given their theological mix, can find ways to legitimately embrace the idea of transhuman and even posthuman life.[37] We will now briefly review some of those possible ways forward.

Monotheistic Religions

In the monotheistic religions, resurrection to new, transformed life is an example of a doctrine we have discussed that can conceivably be understood as a way of interpreting and embracing technological enhancement. In the chapter, "Superintelligence," we saw that discussions about alien life from other worlds could open the door for theological acceptance of posthuman life, as with human life. We are not saying the monotheistic religions will necessarily embrace transhumanism's anthropology. We do contend that these religions have the theological flexibility to do so.

To give another example, the concept of *imago Dei* (image of God), in Judaism and Christianity, can be understood in a way that provides for the acceptance of posthuman beings. As we saw in the introductory chapter and the chapter, "Mind Uploading," humans made in the image of God have been understood as possessing an image of God's rationality, creativity, free will, relationality, and other characteristics. Posthuman beings could conceivably express all of these, just like intelligent life from other planets could, if these beings are understood as being made in the image of God.

Some theologians have explored and even challenged the claim that the image of God is restricted to humans. Professor Trothen writes:

> Joshua Moritz and Celia Deane-Drummond argue that scientific research yields nothing that makes humans qualitatively unique among animals. As a result, it may be, as Deane-Drummond proposes, that animals share in the likeness of God but not the imago Dei as that is the realm only of humans and angels.[38] … [Moritz] stands by the argument that, in keeping with Darwinian science, there

[36] Grumett, "Transformation and the End of Enhancement," 39.

[37] With regard to the Christian religion, this sustained argument is made by Calvin Mercer, "A Theological Embrace of Transhuman and Posthuman Beings," *Perspectives on Science and Christian Faith* 72/2 (June 2020) 83-88.

[38] Celia Deane-Drummond, "God's Image and Likeness in Humans and Other Animals: Performative Soul-Making and Graced Nature," *Zygon* 47, no. 4 (December 2012): 934–48.

is no adequate scientific basis on which to claim that humans are a unique species with qualitatively distinct qualities related to freedom, lack of innate instincts, self-consciousness and self-awareness, culture, rationality, or moral behavior.[39]

If one can mount the argument that there is no scientific basis for claims of human qualitative uniqueness, then there seems little scientific reason to think that extremely altered humans would be made any less in God's image.[40]

Created Co-Creators

In the monotheistic religions especially, God as creator, separate from—though connected to—the created order, is a central theme. In the early stories of the Bible, following the initial creation of the world, the creative activity of God continues in many and varied ways, all the way to the end-time "new heaven and a new earth."[41] The idea of humans as created co-creators is a controversial notion. Those suspicious of technology will charge that humans are "playing God," by engaging in domains and activities of creation that should be reserved only for God.[42]

For those who deny the charge of playing God, or who see the charge of playing God as ambiguous, the created co-creator proposal provides a way to theologically interpret and potentially embrace transhumans, posthumans, and superintelligence. Ted Peters argues that creating, including creating new technology, can be "playing human" as God intended.[43] Since people are created in the image of God, they have a duty to create for the good. God is still at work, along with God's created co-creators, bringing into existence enhanced and new life forms using technology. In this framework, technology can be a means of God's grace, creativity, and salvific activity.

The following reflection, in a Canadian Council of Churches resource and written by Professor Trothen, affirms technology, but in the context of caution that these powerful tools be used wisely.

> Humans, as made in the likeness of God, have been given creativity to be used in divine service. Theologian Philip Hefner's proposal of humans as created co-creators helps us to complicate the caution not to "play God" in the realm of science and technology. While we ought to be prudent and aware of our abilities to mess up, we also have extraordinary capacity to improve life with God's help. The *imago Dei* suggests a divine mandate to create for the good. This is a risky venture requiring humility and some audacity.[44]

[39] Joshua M. Moritz, "Evolution, the End of Human Uniqueness, and the Election of the Imago Dei," *Theology and Science* 9, no. 3 (2011): 312–313.

[40] Tracy J. Trothen, *Winning the Race? Religion, Hope, and Reshaping the Sport Enhancement Debate* (Macon, GA: Mercer University Press, 2015), 176.

[41] In the New Testament, see Revelation 21:1 and 2 Peter 3:13.

[42] "Playing God" is the title of Ted Peters' book, *Playing God: Genetic Determinism and Human Freedom* (New York: Routledge, 1996).

[43] *Playing God.*

[44] Tracy J. Trothen, "Christian Anthropology: Doing Good Through Science and Technology," in *Technology and the Image of God: A Canadian Conversation* (Canadian Council of Churches, Faith and Life Sciences, 2017), 18–19.

Human Fallibility

A related and basic theological anthropology question is, are we mostly good or mostly bad? More specifically, can we be trusted with powerful technology? As we have seen in other chapters, conservatives and liberals tend to articulate different positions on this question, with conservatives emphasizing human depravity and liberals emphasizing human goodness and potential. Earlier in this chapter, we considered some history of how this has played out with regard to the impact of two world wars and terrorism on the (perhaps naïve) optimism of the liberal social gospel movement.

As with most such theological questions, the answer is likely not to be binary—humans are neither all good nor all bad. For various genetic, psychological, social, and spiritual reasons, some people may have stronger tendencies toward evil doing than others. But no single person is all devil or all angel. We all have our "better angels," although we do not always listen to those angels. To put a finer point on the question, it is not so much a matter of whether a particular human being works for good or evil in the world. It is a broader question about the direction of the human race. While this is a theological question, it is not only theological. Most people probably have some sense, whether conscious or not, about the direction in which they think humanity is headed.

The traditions and doctrines of monotheistic religions recognize human failing and sinfulness. The biblical Garden of Eden story[45] is soon followed by a story of Cain killing Able[46] and a devastating flood[47] brought on by human wickedness. The potential of God working through and with humans to use technology for good must be tempered by a recognition of the selfishness, greed, and hatred that also issue in genocide, rape, and a host of other evils.

Hefner was very deliberate to name us humans as *created* co-creators. He makes clear that we are created partly in order to caution against hubris. Hubris is extreme pride and shows up as a will-to-power, creating all kinds of mischief. However, feminist and other liberation theologians have critiqued the equation of sin with pride. Will-to-power is not a danger for many marginalized people. Their danger is more in failing to claim the "co-creator" part of the phrase by being too humble or voiceless. Many black, feminist, and post-colonial theologians have pointed out the danger of giving up one's voice and power in the face of oppressive power structures and attitudes. We need under-represented voices at the table co-designing technology and in the public square.

Unless they are restrained or reenvisioned, powerful technologies can add to power and give power. These technologies can also take away power when they are not co-designed, and they can fail to represent the knowledge, needs, and dreams of not just the powerful but of those on the social margins. Power in

[45] Genesis, chapter 2.
[46] Genesis, chapter 4.
[47] Genesis, chapters 6–9.

itself is not bad. It is power-over that is to be avoided, in a commitment to mutual relationship and the flourishing of all life, as taught by all religions.

Karmic Religions
While there has been less reflection on these issues as they relate to the karmic traditions, it does seem the doctrine of reincarnation can readily accommodate transhumanism's pliable human nature. Based on a being's karmic narrative, the next incarnation can come at various human levels (i.e., castes), but the vision of the karmic religions goes far beyond the human plane. It is believed the soul can transmigrate to levels lower than human or higher. In Buddhism, for example, the six realms of rebirth can include heavenly, demigod, human, animals, ghosts, and residents of hell.

In scenarios with these kinds of possibilities, the monotheistic concerns about embodiment, personhood, and identity are not so much at play. The law of karma applies, whether it is soul in Hinduism or the more complicated Buddhist notion of no-self. Future reincarnations are not dependent on continuity of embodiment, personhood, or identity, at least in the ways these are understood in the monotheistic religions.

Soteriology

Soter is Greek for salvation. Every religion identifies what it considers to be the basic human predicament. For Hinduism, it is ignorance of our true, divine nature; for Buddhism, suffering; for Christianity, sin that separates humans from God. A religion's solution to the basic human predicament is that religion's soteriology, its concept of salvation.

As we have seen, there are some fundamental differences in the theological outlooks of the monotheistic and karmic religions. However, one soteriological idea in certain monotheistic traditions seems quite compatible with that of pantheism, as expressed in Hinduism and other karmic traditions. It is also an idea aligning well with the transhumanist vision of transcending our current human situation. In the chapter, "Transhumanism, the Posthuman, and the Religions," we presented the doctrine of divinization (*theosis*), that is, becoming divine. We see this teaching especially in the Eastern Orthodox branch of Christianity. We also encounter it in the Church of Latter Day Saints, commonly referred to as Mormonism.

A pantheistic religion like Hinduism is more direct and extensive in its assertion that we are divine. The basic human predicament is that we are ignorant of our true, divine nature. We are God—not "a" god, but god in the sense that there is only divinity and therefore all are divine. Eastern Orthodoxy and Mormonism do not have the same doctrinal framework as Hinduism, but there is some family resemblance in that the end result of the sanctification process in these Christian traditions is that we become divine.

In Christianity, sanctification means to make holy, usually through some combination of God's grace and human works. Most Christian traditions stop short of saying that we are to become divine but do encourage followers to work *toward* a realization of God's divine model in Jesus. In Hinduism, salvation means realizing oneness with divine reality. Radical human enhancement that moves us in the direction of omnibenevolence, omniscience, and omnipotence can be understood as serving an ultimate divinization goal as well.

Eschatology

Eschaton is the Greek for "end." So, we come now to the end, the end (or nearly so) of the textbook and to the doctrine of the end-time. How will it all wind up, and is it going to be a heavenly or hellish outcome or somewhere in between?

Earlier in this chapter, we described a right-wing Christian approach to eschatology exhibited in the techno-apocalyptic writing and preaching now gaining steam as a reaction to transhumanist programs. Christian eschatologies can range from this conservative version to more moderate or liberal visions that interpret end-time language (e.g., Antichrist, Armageddon) as symbols of good and evil and the struggle between them, wherever and whenever that occurs.

Transhumanists have an eschatology, a vision of where things are going, although they do not usually clothe it in religious terminology. Ray Kurzweil is an exception with his vision of the universe as becoming saturated with intelligence and waking up. The progressive theological vision of Teilhard de Chardin, discussed earlier in this chapter, can be interpreted so as to incorporate Kurzweil's vision of the Singularity.

As we proposed earlier, the religions have at their command powerful, performative symbols and doctrines by which to interpret and shape technology in ways that reflect religion's values. Whether and how extensively the religions intentionally use their powerful language is yet to be seen. The future of religion and the welfare of society in general depends in part on how the religions address radical human enhancement in the coming years.

Final Ethical Reflections…For Now

Some scientists are very thoughtful about the topics in this textbook, and some of them have advised that we avoid the more radical enhancement efforts. In general, however, scientists are unrestrained in their pursuit of research that will lead to the increased engineering of human beings and more deliberate alteration of the species. More often than not, they are specialists working hard in their labs on some tiny piece of the puzzle that fits into a larger picture they may not clearly see. In general, while scientists certainly should have a voice in deliberations about the ethics of scientific research, it is not solely the responsibility of the scientific community to determine the appropriate aims and limits of research. Scientists should not make decisions on their own with which the rest of humanity will have to live.

Likewise, neither politicians nor the courts should be saddled with these ethical decisions, and the leading politicians to date have remained mostly silent on the questions. Business, including a host of technology start-up companies, is a major player in AI and other developments that can bring radical enhancements. While there is nothing wrong with successful business and making money, it is folly to trust the direction of research, the ethics of these powerful technological programs, and the future of humanity to the corporate bottom line.

A few ethicists and theologians are discussing some of these questions, but thus far they are talking mostly to each other. Thoughtful people of faith, of all religious persuasions, should add their voices to the public square conversation about the technological future and do so now, or society may stumble blindly forward into an unexamined or scary future guided by ambitious scientists, self-serving politicians, and wealthy investors.

An important way for the religions to add their collective voices and wisdom to the conversation about transhuman and posthuman possibilities is to carefully process the relevant technologies and the enhancement programs through the three ways of framing the issue introduced in the chapter, "Radical Human Enhancement and Ethics." The new world coming will align with the best values of the religions, and benefit society, if that world values careful reflection on what it means to be better, choice in the context of relational autonomy, and justice.

Postscript—Coming Full Circle

We arrive at the end of our textbook, and we invite you to consider again how you now describe yourself with regard to radical human enhancement. With regard to radical biohacking, are you:

- Totally skeptical and opposed.
- Open to the possibilities but needing to be convinced about particular enhancements.
- An eager transhumanist, i.e., an eager advocate for radical enhancement.
- Unsure.

As you may recall, we asked this same question at the beginning of this textbook. Is your answer different from your response at the beginning? What has changed and why? Regardless of your answer, we hope you agree it is incumbent upon students of religion, faith communities, and all people to be educated about these issues, in order to thoughtfully contribute to the conversation about them.

Questions for Discussion

1. Do religions have a role to play in the development of radical enhancement technologies? What do you think this role should be, if any? Why?
2. What do you think are the dangers of emergent technologies that the religions need to address?
3. Do you think human beings are basically good or evil? On what basis do you give that answer? What is the significance of this question for enhancement technology?
4. Who do you think will finally decide the direction of enhancement research: governments, the free market, ethicists, religions, or some other organization or entity? Which do you think would be best to decide and why? Which do you think would be worst and why?
5. How likely do you think it is that terrorists or rogue countries will acquire enhancement technologies? What will be the result if they do?
6. Do you think language shapes reality? If so, what kind of language and how?
7. If you were religious, would you consider joining a transhumanist organization in your religion? Why or why not?

Glossary

Affect Feeling, emotion, and mood.

Affective enhancement Enhancing affect through biohacking in various ways.

Alcor Life Extension Foundation The largest and most well-known cryonics organization.

Allopathic medicine Science based, conventional Western medicine provided in hospitals by medical doctors and healthcare professionals. Allopathic medicine relies on pharmaceutical drugs, radiation, and surgery to treat symptoms and disease. Allopathic medicine is distinct from homeopathy, naturopathy, traditional indigenous medicine, and other complementary approaches to healthcare.

Anthropology The study of what it means to be human individually and socially. In this textbook we are interested in theological anthropologies, which concern the study of what it means to be human from the perspective of a particular religious tradition. See "theological anthropology."

Apocalyptic A worldview that generally envisions catastrophic end-of-the-world scenarios. See "premillennial dispensationalism."

Artificial General Intelligence See Artificial intelligence.

Artificial intelligence (AI) The development of computer systems to engage in what have been solely human activities including, but not limited to, processing information, solving complex problems, programming other machines, and relational engagement with humans. "Weak" or" narrow AI" performs particular functions with a predefined range, such as playing chess, driving cars, and helping to diagnose patients. "Strong AI" or "Artificial General Intelligence (AGI) is machine intelligence that mimics human intelligence.

Autonomy Widely recognized as a key bioethical principle, the definition of respect for autonomy is increasingly debated. It includes respect for a

C. Mercer, T. J. Trothen, *Religion and the Technological Future*,
https://doi.org/10.1007/978-3-030-62359-3

person's dignity, which is usually understood as including the ability to make choices about oneself. In normative North American culture, extreme individualism has reduced autonomy to the popular notions of individual rights and individual choice. A relational approach to autonomy understands rights and choice as relational concepts having meaning only in the context of each individual's life narrative, which necessarily includes all effects our choices may have on other people and other life.

Bioconservative Someone who takes a conservative stance on human enhancement and other technological issues.

Biohacking Changing our biological selves, including neurology, biochemistry, physiology, and other physicalities through science and technology for enhancement purposes. Radical human enhancements are achieved via biohacking.

BioLogos A Christian organization that illustrates the dialogue/integration model of the relationship between science and religion. It is founded by Francis Collins, a widely respected scientist who headed up the important Human Genome Project, completed in 2003.

Christian Transhumanist Association (CTA) Founded and led by Micah Redding, the organization is pro-enhancement, but its affirmation is modest enough to be inclusive of Christians ranging from ardent advocates to more cautious proponents. The CTA held its first national conference in 2018, featuring a presentation by Aubrey de Grey.

Cloning, therapeutic See "therapeutic cloning."

Co-design The involvement of all stakeholders in decision-making processes about the design, development, and use of technology, including AI. Stakeholders include everyone affected by the technology. Co-design helps ensure that the technology meets diverse needs and is as usable as possible. The principle of co-design recognizes that socially marginalized communities are usually not well represented in the creation of technology.

Cognition Mental processes involved in comprehending and acquiring knowledge. Cognition includes several processes including but not limited to creative thinking, thought processing, attention, memory, problem solving, sensation, perception, interpreting emotions, and processing stimuli.

Cognitive enhancement Making ourselves smarter by biohacking in various ways.

Consent In bioethics, the process of making a choice regarding medical interventions. Informed consent requires disclosure, comprehension, voluntariness, and competence. Disclosure means that all relevant information, including potential harms and benefits and alternative approaches, are provided to the patient. Comprehension means the person must be able to understand the information in terms of language and educational level.

Voluntariness means that the person must not be pressured or coerced in any way. Competency refers to the capacity to consent and draws attention to any limitations caused by dementia, mental health, or other issues.

Conservative, religious Tends to be God-centered, otherworldly, focused on special revelation from God (e.g., scripture), traditional, and dogmatic. See the "Theological Continuum" table in the chapter, "Transhumanism, the Posthuman, and the Religions."

Created co-creators Coined by Christian theologian Phillip Hefner, the notion that human beings are created co-creators with God, one idea that lays groundwork for viewing at least some technology as a potential means of God's work in the world.

CRISPR Stands for Clustered Regularly Interspaced Short Palindromic Repeats. It is a technology that can be used to edit genes, providing radical enhancement possibilities.

Cryonics A Greek term, meaning "icy cold," it refers to a process that preserves the body in a way that it can be theoretically revived in the future when the cure for the illness causing death is available. Likewise, cryonics offers the possibility of accessing future enhancement therapies and technologies.

Cyborg Short for "cybernetic organism," a cyborg is an organism integrated with artificial technology whose operation restores the organism's so-called normal ability or takes it beyond normal functioning.

DARPA The United States' Pentagon's Defense Advanced Research Projects Agency (DARPA), with a 3.5 billion dollar plus annual budget, is heavily involved in human enhancement in the service of more effective soldiers and a stronger military. DARPA has an impressive track record, having helped pioneer, for example, the internet and email. It supported funding for the computer graphics industry, cell phones, "own-the-night" night-vision sensors, advanced fuel cells, weather satellites, the Saturn V rocket that took humans to the moon, and various "stealth" technologies.

Deontological Ethics Sometimes called "principilist ethics," deontological ethics is a normative ethical theory that moral behavior should be based on principles. Examples of principles are beneficence (duty to do good), non-maleficence (duty to do no harm), respect for autonomy (includes respect for individual choice and dignity), justice (includes procedural and distributive justice), veracity (truth-telling), fidelity, and self-care. Principles are derived from virtues we most value. Principles are codes of behavior, more universal than values, and, prima facie principles are believed to be universally applicable.

Digisexuals People who are sexuality and/or romantically attracted to robots or androids.

Dispensationalism See "premillennial dispensationalism."

Distributive justice The socially just distribution of resources. The resources distributed may be tangible (e.g., food, pay, or technology) or intangible (e.g., encouragement, valuing people by engaging with them in conversation).

Dualism As used in this textbook, refers to the idea that (1) the soul is distinct from the body or (2) there are two equally powerful deities, a good god and an evil god, who are in conflict with each other. The ancient worldviews reflected in neo-Platonism and Gnosticism are dualistic to one degree or another.

Eastern Orthodoxy One of the three branches of Christianity, along with Roman Catholicism and Protestantism.

Ecophagy Literally "eating the environment," the term refers to the "grey goo" end-of-the-world as we know it threat, that intelligent nanobots will go off-mission and manipulate matter at the atomic level into a pile of grey goo.

Embodied cognition Cognition as intertwined with the body in its dynamic interaction with the natural and sentient environment. A case can be made for "body memory" that is not confined to the brain.

Emergence theory The view that new realities come into existence from lower level realities, and the new structures and capacities cannot be reduced to the lower levels. For example, the human brain, as a physical organ of the body, consists of neurons, neurotransmitters, blood glucose, oxygen, electrical impulses, chemical reactions, and a host of other parts and processes. From all this we get sensation, perception, and subjective conscious experience. Applied to religion, the idea is that God is emergent from the universe.

End-time Refers to religious views about how the world as we know it will end.

Entheogen From the Greek meaning "full of god" (*entheos*) and "to come into being" (*genesthai*), it refers to the use of hallucinogenic agents in religious contexts.

Epistemology Theory of knowledge and a subfield in philosophy. Christian liberation theology claims that those on the social margins have an epistemological privilege, meaning that the socially marginalized know most about injustice and, therefore, have the most important input in justice work. Epistemology, a key issue in ethics, reveals the location of power.

Eschatology Doctrine of the end-time. *Eschaton* is Greek for "end."

Ethics Addresses the accepted rules, actions, and behaviors in a community or group.

Exotheology Coined by Christian theologian Ted Peters, the term refers to speculation on the theological significance of extraterrestrial life.

Gene drives Genetic engineering technology used to change genes in germline cells, thereby propagating changes through an entire species in a very effective way.

Germline genetic modification technology Genetic modification technologies targeting the genes in germline cells, including sperm, eggs, and embryos, which are passed from generation to generation.

Hedonism A teleological ethical theory that is interested in the maximizing of pleasure. In short, whatever results in the most pleasure for any person or group is warranted, without regard for other potential consequences.

Humanity 2.0 A term referring to enhanced humans and, possibly, posthumans.

Imago Dei Latin for "image of God," a term used in the creation story in the Bible book of Genesis. The term has elicited much reflection and interpretation in Christian theological writings.

Incarnation The Christian doctrine that God was "fleshed out" in a human being, Jesus Christ, leading to the doctrine that Jesus Christ was fully God and fully human.

Information-theoretic criterion The criterion the cryonics industry uses for death. With this criterion death occurs when the biological structures that encode memory and personality are destroyed, such that this critical information cannot be recovered.

Karma, **Law of** Sanskrit for "action" or "deed" and referring to the idea that one's actions (or inactions) influence, even determine, the status of the lives into which a soul is born. Science has taught us the law of cause and effect. Every event in the physical universe has a cause, and every cause has determinate effects. The law of karma extends cause and effect to the moral and spiritual realms. Moral, spiritually good actions in this life promote rebirth into a next life that is closer to liberation.

Karmic religion Hinduism, Buddhism, Daoism, Jainism, and Sikhism, although each religion understands karma and rebirth in varying ways.

Liberal, religious Tends to be human-centered, this-worldly, rational, revisionist, and pragmatic. Liberal, as used in this textbook, does not necessarily mean individualistic. See the "Theological Continuum" table in the chapter, "Transhumanism, the Posthuman, and the Religions."

Longevity Dividend The idea that significant public funds are saved if people live healthy longer, because most medical dollars are usually consumed in the last years of life, during failing health. The call to action for public officials is to invest in "anti-aging research," as opposed to mere "disease research."

Longevity Escape Velocity (LEV) A concept championed by Aubrey de Grey and Ray Kurzweil. Also called "actuarial escape velocity," LEV refers to a coming future time in which we have the know-how to extend our lives long enough for the next scientific breakthrough to come, which will again extend our life expectancy for the next breakthrough. The title to this book by Kurzweil encapsulates the LEV idea, *Fantastic Voyage: Live Long Enough to Live Forever*.

Mindcloning Creating a digital version of the mind. "Mindware" software activates a "mindfile," a digital repository of memories, feelings, and thoughts.

Mind copying See "whole brain emulation."

Mindfile See "mindcloning."

Mind transfer See "whole brain emulation."

Mind uploading See "whole brain emulation."

Mindware See "mindcloning."

Monotheism From the Greek words *mono* for "one" and *theos* or *theoi* for "god" and "gods," the belief in one god, illustrated most clearly in the "Abrahamic faiths" of Judaism, Christianity, and Islam.

Morals Morals are about our personal character and beliefs about right and wrong.

Moral bioenhancement Enhancing moral capacity through biohacking, usually with pharmaceuticals.

Mormon Transhumanist Association (MTA) The oldest pro-transhumanism religious group started in 2006. The MTA is a robust organization with professional leadership and an active schedule of conferences and other programs. Lincoln Cannon is a founder, board member, and former president of the MTA. "Transfigurism" is the theological word the MTA often uses for "transformation."

Morphological freedom Freedom to modify one's body, the right to be as one wishes as long as it does not interfere with anyone else's right to act similarly. Philosophers maycall such a view "ontological libertarianism," while in popular culture it is sometimes known as "shapeshifting."

Nanotechnology A "nano" is one billionth of a meter, the width of about five carbon atoms. Nanotechnology research is active with animals and may eventually produce blood-cell sized, computerized tools called nanobots, capable of manipulating human biology at the cellular level.

Neuropreservation The practice in cryonics of preserving the head only, since the presumption is that revival will occur far enough into the future that the rest of the body can be regrown or provided by robotics.

Nootropics Widely used smart drug supplements, many sold as "natural."

No-self A Buddhist doctrine that human beings have no permanent, unchanging soul or essence. Buddhism does teach reincarnation, giving various explanations of how this relates to the no-self doctrine.

Omega Point Coined by paleontologist and geologist Pierre Teilhard de Chardin (1881–1955), a Christian priest, the term refers to God as the culmination and unification of complexity and consciousness. For Teilhard de Chardin, the process is driven by love, a principle easily affirmed by all religions.

Omnibeneficience See "omnibenevolence."

Omnibenevolence All-good or all-loving, an attribute of some understandings of deity. Also called omnibeneficience.

Omnipotence All-powerful, an attribute of some understandings of deity.

Omniscience All-knowledge, an attribute of some understandings of deity.

Pantheism From *pan* (Greek for "all" or "everything") and *theos* or *theio* (Greek for deity), the doctrine that god is everything, and everything is god. All reality is ultimately the same "thing," i.e., divine.

Personal identity A philosophical issue about what makes a person the same person persisting through time.

Posthuman A future, sentient being that is not human. See also "transhuman."

Practical immortality See "superlongevity."

Principlist ethics See "deontological ethics."

Precautionary principle Moving new therapies and technologies along slowly, paying very careful attention to possible unknown and unintended harmful side effects, and, above all, doing as little harm as possible. See "proactionary principle."

Premillennial dispensationalism A type of modern Christian apocalypticism that envisions horrific battles between God and Satan leading up to a literal 1000-year peaceful reign of Jesus Christ. Prior to Jesus' victory, evil forces will seek to capture the minds of people, and Christians will be pitted against the world, controlled by Satan. The transhumanist agenda is seen as part of the evil world order. See "apocalyptic."

Principles More universal than values and often difficult to interpret, principles are codes of behavior and are directive as they apply to behavior and decisions. Most ethicists do not regard prima facie principles (i.e., duties that are binding) as absolute, meaning that the principles are required to be followed, unless the principles come into conflict with each other. For example, if doing good (i.e., beneficence) and doing no harm (i.e., non-maleficence) are both prima facie principles, then technology with potential for good and harm presents a conundrum. The potential goods and harms must be weighed, and one or both of the principles will need compromising to a degree.

Proactionary principle An approach to radical human enhancement that advocates for the development of life-improving and even life-prolonging technologies, despite risks, sometimes including what might be considered significant risks. The term was apparently coined by Max More, an early transhumanist, as a way of countering the prevailing precautionary principle. Transhumanists generally favor a proactionary stance. See "precautionary principle."

Procedural justice Fairness of the processes that lead to outcomes. Co-design can contribute to procedural justice. See "co-design." Questions such as how much time is allocated to whom and who makes decisions are important to procedural justice. When individuals have a voice in the process or the process involves characteristics such as consistency and fairness, then procedural justice is enhanced. In the healthcare context, procedural justice questions include who receives care, how long must they wait for care, what quality of care do they receive, and what are the roles of the patient, the patient's family, or multi-disciplinary healthcare professionals.

Process theology Based on the work of English mathematician and philosopher Alfred North Whitehead, at a basic level, process thinking begins with the real world, wherein everything changes all the time. Christian theologians, such as John B. Cobb and Charles Hartshorne, influenced by Whitehead, argued that since everything changes, God does as well.

Protestantism One of the three branches of Christianity, along with Roman Catholicism and Eastern Orthodoxy.

Prolongevity See "superlongevity."

Radical human enhancement Using therapies and technologies now available or anticipated in the future to biohack ourselves in physical, cognitive, affective, moral, and spiritual ways that take us beyond normal functioning in these five categories. Some people prefer the word "extreme" instead of "radical."

Radical life extension See "superlongevity."

Regenerative medicine A broad field that includes tissue engineering and is pushing forward on repairing, improving, and replacing damaged or diseased organs and tissues in our bodies.

Reincarnation Also called rebirth or transmigration of the soul, the doctrine teaches that in the world of illusion the human soul transmigrates through successive lives until it achieves liberation from the cycle. *Samsara* is Sanskrit for "wandering" or "going through."

Religion 2.0 Religion in the transhuman/posthuman era.

Resurrection, transformational Distinct from the resuscitation of a dead body, this is a raising to eternal life with God, with new abilities and possibilities.

Roman Catholicism One of the three branches of Christianity, along with Eastern Orthodoxy and Protestantism.

Self-reflexivity In the service of good ethical analysis, self-reflexivity is reflecting on one's own story from different standpoints in ways that consider systemic privileges and barriers of intersecting social systems like sexism and racism.

Singularity The predicted dramatic, sudden future break in human history when general human intelligence is surpassed. The Singularity could bring extraordinary benefits, such as increased abilities to solve seemingly intractable problems, and it could also bring apocalyptic dangers.

Social justice Concern for equity (e.g., just fairness), particularly for the socially vulnerable and the socially marginalized. Social justice is about the protection and empowering of those with less power due to racialization, socio-economic disadvantage, ageism, disability, sexism, gender, and/or sexual identity discrimination, or any other similar factor.

Soma pneumatikon See "spiritual body."

Spiritual body Translation of the Greek *soma pneumatikon* (literally "body spiritual"), this term refers to the transformed life as discussed in the Christian scriptures by the writer Paul, who understood that the body, after death and in the resurrection, would be transformed into being imperishable, glorious, and powerful. While dramatically changed for the age to come, the spiritual body is a clear continuation of the pre-transformed person. See "resurrection, transformational."

Situationalism An ethical theory holding that all moral decisions are particular to the specific situation; there are no overriding norms. Each situation must be understood apart from preconceived conclusions or rules. However, even situationalists usually acknowledge the one overarching rule that love must be maximized.

Somatic genetic modification technology Adding, deleting, or changing genes in specific types of cells (e.g. blood cells, skin cells) in one person and in a way that is not intended to be passed on to progeny.

Soteriology The doctrine of salvation. *Soter* is Greek for salvation.

Sources of authority An ethical concept referring to aspects of our context that shape our values, opinions, and beliefs. These aspects are sources of "authority," because we give them the power of influence. Sources of authority include three broad areas: experience (e.g., school, family, relationships, intuition), tradition and history (e.g., sacred texts, religious practices and teachings, social practices, rituals, historic interpretations), and science and reason (e.g., media, law, research). All views are limited, partial, and perspectival. Since our context changes over time, our sources of authority and, therefore, our values, opinions, and beliefs, can also change over time.

Soul A word interpreted in different ways in the various religions. Generally, in the karmic religions, the soul is the eternal and indestructible basic self that is reborn as human, animal, and heavenly or hellish beings until final liberation from the reincarnation "wheel of rebirth." In some strands of the monotheistic religions, a human being is viewed as a psychosomatic unity of body and soul. Soul is not a distinct and separate "part" of what constitutes a person. However, in some conservative wings, soul is distinct from the body and the object of salvation.

Spirit tech Technological interventions that allegedly provide authentic spiritual experiences.

Spiritual enhancement Enhancing spirituality through biohacking in various ways.

Strong AI See Artificial intelligence.

Superintelligence Machine intelligence that surpasses Atificial General Intelligence, i.e., the typical intelligence of an average human being.

Superlongevity Also called "radical life extension," "extreme longevity," "prolongevity," and "practical immortality," the term refers to radically extending healthy human life indefinitely through any number of biohacking methods. "Practical" is used in "practical immortality," because death could still occur though accidents or some global event like nuclear war.

Technology Applied scientific knowledge. In this textbook we are particularly interested in the application of science that contributes to the biohacking of humans for the purpose of making humans better.

TechPlus Theology Coined for this textbook, the term refers to doing theology in the context of technologically enhanced humans and, possibly, posthumans. Each religion will construct theologies based on that religion's scriptural and theological tradition. And, of course, there are likely to be numerous theological programs in each religion, ranging through the liberal-conservative continuum.

Teleological Ethics theories emphasizing possible and anticipated consequences in decision-making. "*Telos*" is Greek meaning "end" or "goal." Teleological or consequentialist theories are usually thought to include situationalism and utilitarianism. See "situationalism" and "utilitarianism."

Telomeres Caps at the end of DNA strands that protect the chromosomes and become damaged and shortened, thereby hastening death. Telomeres are the subject of much interest regarding superlongevity.

Theistic, non-theistic Adjectives referring, respectively, to believing in god(s) or not. Some religious traditions are non-theistic, such as Theravada, one of the two main branches of Buddhism.

Theodicy Literally "God justice," it is reflection on the question of how one can justify the existence of a supremely wise, powerful, good deity in the face of suffering and evil.

Theology From the Greek word *theos* (god) or *theoi* (gods), the study of religious belief, also called doctrine. Theology proper is "word about God" or study of the divine. We have used the term theology in both senses throughout this textbook.

Theological anthropology A religion's doctrine, or belief, about human beings.

Theological continuum Broad generalizations about liberal and conservative that can be useful in beginning to understand ideologically based movements or trends. A reference table is provided in the chapter, "Transhumanism, the Posthuman, and the Religions."

Theosis Also called "deification," an important theological notion in Eastern Orthodox Christianity well expressed in the often-quoted words of Athanasius of Alexandria (296/298—373 CE), "He [Jesus Christ] was made man that we might be made god." Theosis refers to a view of salvation whereby one is transformed, or divinized, and elevated in some significant way into the life of God. The discussions about (and terminology of) *theosis* are quite technical and complex, perhaps in part due to the effort to stop short of attributing full divinity to a human, which would be heresy in a monotheistic religion. The idea is also affirmed by Mormonism, a Christian movement.

Therapeutic cloning Producing organs and tissues for transplanting into humans.

Therapy—enhancement continuum An approach traditionally used to assess enhancement interventions on a continuum from clearly therapeutic (maintaining or bringing the patient to whatever is considered normal functioning) to clearly enhancement (moving the patient to above normal functioning with little or no therapeutic value). In this textbook, religious integrity is proposed as a dividing point (instead of the fraught concept of normal) between acceptable and unacceptable interventions.

Transcendence Qualities and feelings of mystery and ineffability. Transcendence also involves overcoming limitations and reaching for "something more." Religion and transhumanism share a desire for transcendence but may understand the meaning of "something more" differently.

Transhuman Narrowly conceived, the term transhuman refers to that which is still human, albeit greatly enhanced. However, the term is sometimes used more loosely, in the broader sense of (1) sentient beings that develop from humans but to such a degree that they are no longer human in any real sense, and (2) sentient beings that develop apart from humans.

Transhumanism As posted on the American Academy of Religion website: "'Transhumanism' or 'human enhancement' refers to an intellectual and cultural movement that advocates the use of a variety of emerging technologies. The convergence of these technologies may make it possible to take control of human evolution, providing for the enhancement of human mental and physical abilities deemed desirable and the amelioration of aspects of the human condition regarded as undesirable. These enhancements include the radical extension of healthy human life." See also "transhuman" and "posthuman."

Utilitarianism A teleological ethical theory that looks to maximize the greatest good for the greatest number. Utilitarians agree our overall aim in evaluating actions should be to create the best results possible, but they differ about how to do that. Act utilitarians focus on the effects of individual actions; they take a case-by-case approach. Rule utilitarians accentuate moral rules and emphasize the general effects (consequences) of actions, such as killing or stealing.

Values Those things held most important and, as such, that give direction in life. Values are sets of beliefs about subjective traits and ideals, with examples being relationships, family, health, long life, faith, caring, justice, education, ambition, adventure, and wealth.

Virtue Ethics An ethical theory that asks not what should I do but what sort of person should I be. Highlighting character, the key question is, "What would the most virtuous person we can think of do in a similar situation?" For Christians, this person might be Jesus. For Buddhists, the virtuous person might be the Buddha. Some people think beyond the iconic "founders" of religions to great saints or other figures. Virtues are qualities we deem to be morally good or desirable in people and might include prudence, self-control, generosity, and kindness.

Vitrification A chemical ice-free preservation technique used by cryonics in an attempt at structural preservation of the brain for later "thawing."

Weak AI See Artificial intelligence.

Whole brain emulation Technical term for what is commonly called "mind uploading" and sometimes "mind copying" or "mind transfer." It refers to copying the information (i.e., memory and personality) in the brain into a digital substrate (part of a computer). Mind uploading is the stuff of science fiction that has made its way into serious conversation about transhumanist possibilities. Calling this procedure "mind" uploading treads into very complex questions about the definition of mind.

References

Abbott, Alison. 2019. First Hint That Body's 'Biological Age' Can Be Reversed. *Nature* (September 5). https://www.nature.com/articles/d41586-019-02638-w.

Abelman, D.D. 2017. Mitigating Risks of Students Use of Study Drugs Through Understanding Motivations for Use and Applying Harm Reduction Theory: A Literature Review. *Harm Reduction Journal* 14 (68). https://doi.org/10.1186/s12954-017-0194-6.

Accolla, Ettore A., and Xlauda Pollo. 2019. Mood Effects After Deep Brain Stimulation for Parkinson's Disease: An Update. *Frontiers in Neurol*ogy (June 14). Accessed June 5, 2020. https://doi.org/10.3389/fneur.2019.00617.

Achenbach, Joel. 2015. A Harvard Professor Says He Can Cure Aging, But Is That a Good Idea? *Washington Post* (December 2). https://www.washingtonpost.com/news/achenblog/wp/2015/12/02/professor-george-church-says-he-can-reverse-the-aging-process/.

Adam, Aalia. 2018. Meet Pepper—Canada's First Emotionally Sensitive Robot for Sick Kids. Interview with *Global News*. (May 7). https://globalnews.ca/news/4180025/pepper-Canada-robot/.

Adrien, C. 2017. We Live in a Death Denying Culture. That's a Problem (Blog). 1800 Hospice. Accessed https://www.1800hospice.com/blog/live-death-denying-culture-thats-problem/.

"Affectiva." n.d. affectiva.com.

AFP. n.d. US Trial Shows 3 Cancer Patients Had Their Genomes Altered Safely by CRISPR. *ScienceAlert*. Accessed February 8, 2020. https://www.sciencealert.com/researchers-genetically-alter-the-immune-system-of-cancer-patients-without-side-effect.

Agar, Nicholas. 2011. Kurzweil and Uploading: Just Say No! *Journal of Evolution and Technology* 22 (1): 23–36.

Alcaraz, Aleksandra Lukaszewicz. 2021. *Are Cyborgs Persons? An Account of Futurist Ethics. In Palgrave Studies in the Future of Humanity and Its Successors*. Calvin Mercer and Steve Fuller, Series Co-editors. New York: Palgrave Macmillan.

Alcor Life Extension. n.d. Frequently Asked Questions. https://alcor.org/FAQs/faq03.html#world.

C. Mercer, T. J. Trothen, *Religion and the Technological Future*,
https://doi.org/10.1007/978-3-030-62359-3

Alcor Life Extension Foundation. n.d. "Alcor Membership Statistics." "The Alcor Team." "CEO Statement on Membership Statistics." "Cryonics Myths." "Frequently Asked Questions." "Scientists' Cryonics FAQ." "Scientists Open Letter on Cryonics." "What is Cryonics." http://www.alcor.org.

Al-Kulayni. n.d. Vol. 1, 39. Accessed May 25, 2020. https://blogs.scientificamerican.com/observations/the-runaway-train-of-cognitive-enhancement/.

Al-Rodhan, Nayef. 2019. The Runaway Train of Cognitive Enhancement. *Scientific American* (December 9). https://blogs.scientificamerican.com/observations/the-runaway-train-of-cognitive-enhancement/.

Ançel, G. 2006. Developing Empathy in Nurses: An Inservice Training Program. *Archives of Psychiatric Nursing* 20 (6): 249–257.

Armstrong, David. 1968. *A Materialist Theory of the Mind.* London: Routledge & Kegan Paul.

Arnold, Edwin, Trans. 2015. *The Bhagavad Gita.* Kansas City: Scholar's Choice.

Asimov, Isaac. 1942. Runaround. *Astounding Science Fiction.* March.

"Astrotheology & Astroethics." *Theology and Science* 16/4 (2018).

Athanasius of Alexandria. 2011. *On the Incarnation of the Word.* Popular Patristics Series 44. Trans. John Behr. Yonkers: St. Vladimir's Seminary Press.

Ayan, Steve. n.d. The Brain's Autopilot Mechanism Steers Consciousness. *Scientific American* (12/19/18). https://www.scientificamerican.com/article/the-brains-autopilot-mechanism-steers-consciousness/.

Barrat, James. 2013. *Our Final Invention: Artificial Intelligence and the End of the Human Era.* New York: St. Martin's Press.

Bassett, Debra. *The Creation and Inheritance of Digital Afterlives: You Only Live Twice.* In Palgrave Studies in the Future of Humanity and Its Successors, series coeditors Calvin Mercer and Steve Fuller. New York: Palgrave Macmillan, forthcoming.

Batt-Rawden, S.A., M.S. Chisholm, B. Anton, and T.E. Flickinger. 2013. Teaching Empathy to Medical Students: An Updates, Systematic Review. *Academic Medicine* 88 (8): 1171–1177.

Bedell, George C., Leo Sandon Jr., and Charles T. Wellborn. 1975. *Religion in America.* New York: Macmillan.

Benedet, T., et al. 2017. Transcriptional Repressor DREAM Regulates Trigeminal Noxious Perception. *Journal of Neurochemistry* 141 (4): 544–552.

"Biohax International: Turning the Internet of Things into the Internet of Us." n.d.. https://www.f6s.com/biohaxinternational.

BioLogos. n.d. Life Beyond Earth: What Would It Mean for Christians? https://biologos.org/resources/life-beyond-earth-what-would-it-mean-for-christians.

Bird, Hilary. 2017. AI Innovation Could Cause an Ethical Conundrum for Organized Religion. *VB.* October 16. https://venturebeat.com/2017/10/16/ai-innovation-could-cause-an-ethical-conundrum-for-organized-religion/.

Blackford, Russell. 2014. Introduction II: Bring on the Machines. In *Intelligence Unbound: The Future of Uploaded and Machine Minds*, ed. Russell Blackford and Damien Broderick, 11–25. West Sussex, UK: Wiley Blackwell.

Blackford, Russell, and Damien Broderick, eds. 2014. *Intelligence Unbound: The Future of Uploaded and Machine Minds.* West Sussex, UK: Wiley Blackwell.

Bostrom, Nick. 2005. The Fable of the Dragon-Tyrant. *Journal of Medical Ethics* 31 (5).

———. 2014. *Superintelligence: Paths, Dangers, Strategies.* Oxford: Oxford University Press.

———. 2016. TED TALK: What Happens When Our Computers Get Smarter Than We Are? April 27. https://www.youtube.com/watch?v=MnT1xgZgkpk.

Braga, Adriana, and Robert K. Logan. 2020. The Emperor of Strong AI Has No Clothes: Limits to Artificial Intelligence. In *AI and the Singularity: A Fallacy or Great Opportunity?* ed. Robert K. Logan and Adriana Braga, 5–25. Basel: MDPI.

Brashear, R.P. (Director). 2013. Fixed: The Science/Fiction of Human Enhancement. *New Day Films.* http://www.fixedthemovie.com.

Braxton, Donald M. 2015. Does Transhumanism Face an Uncanny Valley Among the Religious? In *Religion and Transhumanism: The Unknown Future of Human Enhancement*, ed. Calvin Mercer and Tracy J. Trothen, 331–350. Westport, CT: Praeger.

Bregman, Lucy. 1999. *Beyond Silence and Denial—Death and Dying Reconsidered.* Louisville, KY: Westminster John Knox Press.

Brody, H., and A. Macdonald. 2013. Religion and Bioethics: Toward an Expanded Understanding. *Theoretical Medicine and Bioethics* 34: 133–145.

Bromwich, Jonah Engel. 2019. Maybe We Are Wired to Beat Up Machines. *The New York Times, SundayStyles* (January 20): 1, 8.

Brooker, Katrina. 2015. Google Ventures and the Search for Immortality. *Bloomberg* (March 9). www.bloomberg.com/news/articles/2015-03-09/google-ventures-bill-maris-investing-in-idea-of-living-to-500.

Brooks, Rodney A. 2002. *Flesh and Machines: How Robots Will Change Us.* New York: Vintage Books.

Brown, Mick. 2020. The Billionaire Tech Entrepreneur on a Mission to Defeat Death. *The Telegraph* (September 19).

Burdett, Michael. 2015. The Religion of Technology: Transhumanism and the Myth of Progress. In *Religion and Transhumanism: The Unknown Future of Human Enhancement*, ed. Calvin Mercer and Tracy Trothen, 131–148. Westport, CT: Praeger.

Cabrera, Laura Y., Emily L. Evans, and Roy H. Hamilton. 2014. Ethics of the Electrified Mind: Defining Issues and Perspectives on the Principled Use of Brain Stimulation in Medical Research and Clinical Care. *Brain Topography* 27: 33–45.

"Can Google Solve Death?" 2013. *Time* (September 30).

"Can Science Understand Everything? NASA Scientists Attempt to Answer the Question." n.d.. https://aeon.co/videos/can-science-understand-everything-nasa-scientists-attempt-to-answer-the-question.

Cannon, Lincoln. 2017. Mormonism Mandates Transhumanism. In *Religion and Human Enhancement: Death, Values, and Morality*, ed. Tracy J. Trothen and Calvin Mercer, 49–66. In *Palgrave Studies in the Future of Humanity and Its Successors*, Series Editors Calvin Mercer and Steve Fuller. New York: Palgrave Macmillan. Republished in Paperback, 2019.

———. n.d. Mormon Transhumanism. https://www.youtube.com/watch?v=AeyJbROo-Pw.

Carboncopies. n.d.. https://carboncopies.org/.

Cellan-Jones, Rory. 2014. Stephen Hawking Warns Artificial Intelligence Could End Mankind. *BBC News.* December 12. https://www.bbc.com/news/technology-30290540.

Chalmers, David. 2010. The Singularity: A Philosophical Analysis. *The Journal of Consciousness Studies* 17: 7–65. http://consc.net/papers/singularity.pdf.

Chalmers, David J. 2014. Uploading: A Philosophical Analysis. In *Intelligence Unbound: The Future of Uploaded and Machine Minds*, ed. Russell Blackford and Damien Broderick, 102–119. West Sussex, UK: Wiley Blackwell.

de Chardin, Pierre Teilhard. 2003a. *The Human Phenomenon*. Translated by Sarah Appleton-Weber. Portland: Sussex Academic Press (First Published Posthumously 1955).

———. 2003b. *The Divine Milieu*. Translated by Siôn Cowell. Portland: Sussex Academic Press (First Published Posthumously 1957).

Cheron, G., et al. 2016. Brain Oscillations in Sport: Toward EEG Biomarkers of Performance. *Frontiers in Psychology* 7 (246). https://doi.org/10.3389/fpsyg.2016.00246.

Cheshire, William P. 2018. Ethical Implications of Transcranial Magnetic Stimulation for Personal Identity. *Ethics and Medicine: An International Journal of Bioethics* 34 (3): 135–145.

Chorost, Michael. 2005. *Rebuilt: How Becoming Part Computer Made Me More Human*. New York: Houghton Mifflin.

Christian Transhumanist Association. n.d.. https://www.christiantranshumanism.org/.

Chu, Ted. 2014. *Human Purpose and Transhuman Potential: A Cosmic Vision for Our Future Evolution*. San Rafael, CA: Origin Press.

Church of Perpetual Life. n.d.. https://www.churchofperpetuallife.org/.

Clark, Andy. 1997. *Being There: Putting Brain, Body, and World Together Again*. Cambridge, MA: MIT Press.

Clayton, Philip. 2004. *Mind and Emergence: From Quantum to Consciousness*. Oxford: Oxford University.

Cole-Turner, Ron. 2009. Extreme Longevity Research: A Progressive Protestant Principle. In *Religion and the Implications of Radical Life Extension*, ed. Derek Maher and Calvin Mercer. New York: Palgrave Macmillan. Republished in Paperback, 2014.

———. 2015a. Spiritual Enhancement. In *Religion and Transhumanism: The Unknown Future of Human Enhancement*, ed. Calvin Mercer and Tracy J. Trothen, 369–384. Westport, CT: Praeger.

———. 2015b. Spiritual Enhancement. In *Religion and Transhumanism: The Unknown Future of Human Enhancement*, ed. Calvin Mercer and Tracy J. Trothen, 369–383. Santa Barbara: Praeger.

———. 2016. *The End of Adam and Eve: Theology and the Science of Human Origins*. Pittsburgh, PA: TheologyPlus Publishing.

———. 2019. Christian Transhumanism. In *Religion and Human Enhancement: Death, Values, and Morality*, ed. Tracy J. Trothen and Calvin Mercer, 35–48. In *Palgrave Studies in the Future of Humanity and Its Successors*, Series Edited by Calvin Mercer and Steve Fuller. New York: Palgrave Macmillan, 2017; Paperback.

Collins, Francis. 2019. Gene Therapy Shows Promise Repairing Brain Tissue Damaged by Stroke. *NIH Director's Blog* (September 24). https://directorsblog.nih.gov/2019/09/24/gene-therapy-shows-promise-repairing-brain-tissue-damaged-by-stroke/.

Corabi, Joseph, and Susan Schneider. 2014. If You Upload, Will You Survive? In *Intelligence Unbound: The Future of Uploaded and Machine Minds*, ed. Russell Blackford and Damien Broderick, 131–145. West Sussex, UK: Wiley Blackwell.

Cryonics Institute—Technology for Life. n.d.. https://www.cryonics.org/.

De Melo-Martin, Inmaculada, and Arleen Salles. 2015. Moral Bioenhancement: Much Ado about Nothing? *Bioethics* 29: 223–232.

Deane-Drummond, Celia. 2012. God's Image and Likeness in Humans and Other Animals: Performative Soul-Making and Graced Nature. *Zygon* 47 (4): 934–948.

DeBaets, Amy Michelle. 2015. Rapture of the Geeks: Singularitarianism, Feminism, and the Yearning for Transcendence. In *Religion and Transhumanism: The Unknown Future of Human Enhancement*, ed. Calvin Mercer and Tracy J. Trothen, 181–198. Santa Barbara, CA: Praeger.

Detweiler, Craig. 2013. *iGods: How Technology Shapes Our Spiritual and Social Lives and Selfies: Searching for the Image of God in a Digital Age*. Ada, MI: Brazos Press.

———. 2018. *Selfies: Searching for the Image of God in a Digital Age*. Grand Rapids: Baker.

Dinkins, Stephanie, and Charlton McIlwain. 2018. Coding While Black: Artificial Intelligence, Computing, and Data in a Racialized World. *Initiatives Emerging Leaders Program Blog Humanizing Data Review*. New York University (March 12). https://urbandemos.nyu.edu/2018/03/12/coding-while-black-artificial-intelligence-computing-and-data-in-a-racialized-world/.

Doehring, Carrie. 2015. *The Practice of Pastoral Care–A Postmodern Approach*. Louisville. KY: Westminster/John Knox Press, Revised and Expanded Edition.

Dorff, Elliot N. 2014. Becoming Yet More Like God: A Jewish Perspective on Radical Life Extension. In *Religion and the Implications of Radical Life Extension*, Co-edited by Derek Maher and Calvin Mercer, 64–74. New York: Palgrave Macmillan, 2009; Republished in Paperback.

Douglas, T. 2008. Moral Enhancement. *Journal of Applied Philosophy* 25: 228–245.

Dressler, M., et al. 2019. Hacking the Brain: Dimensions of Cognitive Enhancement. *ACS Chemical Neuroscience* 10 (3): 1137–1148. https://doi.org/10.1021/acschemneuro.8b00571.

Drexler, K. Eric. 1986. *Engines of Creation: The Coming Era of Nanotechnology*. New York: Doubleday.

Dreyfuss, Hubert, and Sean Dorrance Kelly. 2011. *All Things Shining: Reading the Western Classics to Find Meaning in a Secular Age*. New York: Free Press.

Durkheim, Émile. 1965. *The Elementary Forms of the Religious Life*. New York: Free Press.

Ertz, Susan. 1943. *Anger in the Sky*. New York: Literary Classics.

Estes, Douglas. 2018. *Braving the Future: Christian Faith in a World of Limitless Tech*. Harrisonburg, VA: Herald Press.

Ettinger, Robert C.W. 1962. *The Prospect of Immortality*. Clinton Township, MI: The Cryonics Institute.

"Euthanasia & Physician-Assisted Suicide (PAS) around the World." n.d. Britannica ProCon (2/26/20). https://euthanasia.procon.org/euthanasia-physician-assisted-suicide-pas-around-the-world/.

Faith and Life Sciences Reference Group, The Canadian Council of Churches. 2017. *Technology and the Image of God: A Canadian Conversation*. Toronto: Canadian Council of Churches, Faith and Life Sciences.

Ferriss, Adam. 2018. AI Is My Shepherd. *Wired* 26 (03).

"Fight Aging!" n.d.. https://www.fightaging.org/.

Fisher, Matthew Zaro. 2015. More Human Than the Human? Toward a 'Transhumanist' Christian Theological Anthropology. In *Religion and Transhumanism: The Unknown Future of Human Enhancement*, ed. Calvin Mercer and Tracy Trothen, 23–38. Westport, CT: Praeger.

Fiske, Amelia, Peter Henningsen, and Alena Buyx. 2019. Your Therapist Will See You Now: Ethical Implications of Embodied Artificial Intelligence in Psychiatry, Psychology, and Psychotherapy. *Journal of Medical Internet Research* 21 (5) https://www.ncbi.nlm.nih.gov/pmc/articles/PMC6532335/. https://doi.org/10.2196/13216.

Foucault, Michel. 1988. *Technologies of the Self.* Boston: University of Massachusetts Press.

Franssen, Maarten, GertJan Lokhorst, and Ibo van de Poel. 2018. Philosophy of Technology. *Stanford Encyclopedia of Philosophy.* Last Updated September 6. https://plato.stanford.edu/entries/technology/.

Frey, Rodney E. 2011. Another Look at Technology and Science. *Journal of Technology Education* 3 (1): 1–12. www.scholar.lib.vt.edu/ejournals/JTE/v3n1/pdf/frey.pdf.

Friend, Tad. 2017. The God Pill—Silicon's Valley's Quest to Live Forever: Can Billions of Dollars' Worth of High-Tech Research Succeed in Making Death Optional? *The New Yorker* (April 3).

Fukuyama, Francis. 2002. *Our Posthuman Future: Consequences of the Biotechnology Revolution.* New York: Farrar, Straus and Giroux.

Fuller, Steve. 2011. *Humanity 2.0: What It Means to Be Human Past, Present and Future.* New York: Palgrave Macmillan.

———. 2019. *Nietzschean Meditations: Untimely Thoughts at the Dawn of the Transhuman Era.* Posthuman Studies 1, ed. Stefan Lorenz Sorgner. Basel: Schwabe Verlag.

Fuller, Steve, and Veronika Lipińska. 2014. *The Proactionary Imperative: A Foundation for Transhumanism.* New York: Palgrave Macmillan.

Garner, Stephen. 2015. Christian Theology and Transhumanism: The 'Created Co-creator and Bioethical Principles. In *Religion and Transhumanism: The Unknown Future of Human Enhancement*, ed. Calvin Mercer and Tracy J. Trothen, 229–243. Westport, CT: Praeger.

Geanacopoulos, Mark. 2004. The Determinants of Lifespan in the Nematode Caenorhabditis Elegans: A Short Primer. *Science Progress* 87 (4): 227–247.

Geraci, Robert M. 2010. *Apocalyptic AI: Visions of Heaven in Robotics, Artificial Intelligence, and Virtual Reality.* Oxford, UK: Oxford University Press.

Geraci, Robert M. 2014. *Virtually Sacred: Myth and Meaning in 'World of Warcraft' and 'Second Life'.* Oxford and New York: Oxford University Press.

Geraci, Robert M. 2018. *Temples of Modernity: Nationalism, Hinduism, and Transhumanism in South Indian Science.* Lanham, Maryland: Lexington Books.

Gill, Sam. 2018. *Religion and Technology into the Future: From Adam to Tomorrow's Eve* (*Studies in Body and Religion*). Lexington Books.

Goertzel, Ben, and Joel Pitt. 2014. Nine Ways to Bias Open-Source Artificial General Intelligence Toward Friendliness. In *Intelligence Unbound: The Future of Uploaded and Machine Minds*, ed. Russell Blackford and Damien Broderick, 61–89. West Sussex: Wiley Blackwell.

Gollner, Adam Leith. 2013. The Immortality Financiers: The Billionaires Who Want to Live Forever. *The Daily Beast* (August 20). http://www.thedailybeast.com/articles/2013/08/20/the-immortality-financiers-the-billionaires-who-want-to-live-forever.html.

Good, Irving John. 1965. Speculations Concerning the First Ultraintelligent Machine. In *Advances in Computers*, ed. Franz L. Alt and Morris Rubinoff, 31–88. New York: Academic Press.

Gould, Stephen Jay. 1999. *Rocks of Ages: Science and Religion in the Fullness of Life.* New York: Ballantine Books.

de Grey, Aubrey. 2013. Undoing Aging: Aubrey De Gray at Tedx Danubia. https://www.youtube.com/watch?v=qMAwnA5WvLc.

———. n.d.-a Aubrey de Grey: A Roadmap to End Aging. https://www.ted.com/talks/aubrey_de_grey_says_we_can_avoid_aging?language=en.

———. n.d.-b Dr. Aubrey de Grey—SENS Research Foundation. Life Extension Advocacy Foundation. https://www.lifespan.io/news/dr-aubrey-de-grey-yuri/.

Grieve, Gregory Price. 2016. *Cyber Zen: Imagining Authentic Buddhist Identity, Community, and Practices in the Virtual World of Second Life.* London: Routledge.

Griffiths, Roland R., Ethan S. Hurwitz, Alan K. Davis, Matthew W. Johnson, and Robert Jesse. 2019. Survey of Subjective 'God Encounter Experiences': Comparisons Among Naturally Occurring Experiences and Those Occasioned by the Classic Psychedelics Psilocybin, LSD, Ayahuasca, or DMT. *PLOS ONE* (April 23). https://doi.org/10.1371/journal.pone.0214377.

Grossman, Lev, and Matt Vella. 2014. How Apple Is Invading Our Bodies. *Time* (September 10). http://time.com/3318655/apple-watch-2/.

Grumett, David. 2011. Transformation and the End of Enhancement. In *Transhumanism and Transcendence: Christian Hope in an Age of Technological Enhancement*, ed. Ron Cole-Turner, 37–49. Washington: Georgetown University.

Habermas, Jürgen. 1971. *Knowledge and Human Interest.* Boston: Beacon Press.

Hall, Douglas John. 1986. *God and Human Suffering.* Minneapolis: Augsburg Publishing House.

Hamzelou, Jessica. 1017. Your Autopilot Mode is Real—Now We Know How the Brain Does It. *New Scientist* (10/23). https://www.newscientist.com/article/2151137-your-autopilot-mode-is-real-now-we-know-how-the-brain-does-it/#ixzz6YPqnV6DL.

Hanrahan, M., and B. Wills. 2015. Makayla's Decision: The Exercise of Indigenous Rights and the Primacy of Allopathic Medicine in Canada. *The Canadian Journal of Native Studies* 35 (2): 207–223.

Hanson, Robin. 2016. *The Age of Em: Work, Love and Life When Robots Rule the Earth.* Oxford: Oxford University.

Harari, Yuval Noah. 2014. *Sapiens: A History of Humankind.* New York: Harper Perennial.

Haraway, Donna. *Simians, Cyborgs and Women: The Reinvention of Nature.* New York: Routledge, 1991.

Harris, John. 2011. Moral Enhancement and Freedom. *Bioethics* 25: 102–111.

———. 2017. How to Welcome New Technologies: Some Comments on the Article by Inmaculada de Melo-Martin. *Cambridge Quarterly of Healthcare Ethics* 26: 166–172. https://doi.org/10.1017/S0963180116000736.

Hauskeller, Michael. 2016. The Art of Misunderstanding Critics: The Case of Ingmar Persson and Julian Savulescu's Defense of Moral Bioenhancement. *Cambridge Quarterly of Healthcare Ethics* 25: 153–161.

Hawking, Stephen. 2014. *Op-ed. Huffington Post*, April.

Hayles, N. Katherine. 1999. *How We Became Posthuman: Virtual Bodies in Cybernetics, Literature, and Informatics.* Chicago: University of Chicago Press.

Hefner, Philip. 1993. *The Human Factor: Evolution, Culture, and Religion.* Minneapolis, MN: Fortress.

Herzfeld, Noreen. 2002. Cybernetic Immortality versus Christian Resurrection. In *Resurrection: Theological and Scientific Assessments*, ed. Ted Peters, Robert John Russell, and Michael Walker, 192–201. Grand Rapids: Eerdmans.

———. 2017. Must We Die? Transhumanism, Religion, and the Fear of Death. In *Religion and Human Enhancement: Death, Values, and Morality*, Coedited by Tracy J. Trothen and Calvin Mercer, 285–99. In *Palgrave Studies in the Future of Humanity and Its Successors*, Series Coeditors Calvin Mercer and Steve Fuller. New York: Palgrave Macmillan.

Holloway, K.R., T.H. Bennett, O. Parry, and C. Gorden. 2013. Misuse of Prescription Drugs on University Campuses: Options for Prevention. *International Review of Law, Computers & Technology* 27: 324–334. https://doi.org/10.1080/13600869.2013.796707.

Holstein, M., J. Parks, and M. Waymack. 2011. *Ethics, Aging, & Society: The Critical Turn*. New York, NY: Springer Publishing Company.

Hopkins, Patrick. 2015. A Salvation Paradox for Transhumanism: Saving You Versus Saving You. In *Religion and Transhumanism: The Unknown Future of Human Enhancement*, ed. Calvin Mercer and Tracy J. Trothen, 71–82. Westport, CT: Praeger.

Horn, Tom, and Nita Horn. 2010. *Forbidden Gates: How Genetics, Robotics, Artificial Intelligence, Synthetic Biology, NanoTechnology, and Human Enhancement Herald the Dawn of Techno-Dimensional Spiritual Warfare*. Crane, Missouri: Defender.

———. 2011. *Pandemonium's Engine: How the Church Age, the Rise of Transhumanism, and the Coming of the Ubermensch (Overman) Herald Satan's Final and Imminent Assault on the Creation of God*. Crane, Missouri: Defender.

Hughes, James. 2004. *Citizen Cyborg: Why Democratic Societies Must Respond to the Redesigned Human of the Future*. Boulder, CO: Westview Press.

———. 2007. The Compatibility of Religious and Transhumanist Views of Metaphysics, Suffering, Virtue and Transcendence in an Enhanced Future. Institute for Ethics and Emerging Technologies (5/8). http://ieet.org/archive/20070326-Hughes-ASU-H+Religion.pdf.

———. 2014. How Conscience Apps and Caring Computers Will Illuminate and Strengthen Human Morality. In *Intelligence Unbound: The Future of Uploaded and Machine Minds*, ed. Russell Blackford and Damien Broderick, 26–34. West Sussex, UK: Wiley Blackwell.

———. 2015. How Moral Is (Moral) Enhancement? Moral Enhancement Requires Multiple Virtues toward a Posthuman Model of Character Development. *Cambridge Quarterly of Healthcare Ethics* 24: 86–95.

———. 2017. Ancient Aspirations Meet the Enlightenment. In *Religion and Human Enhancement: Death, Values, and Morality*, ed. Tracy J. Trothen and Calvin Mercer, 191–212. New York: Palgrave Macmillan.

"Humanity+." n.d.. http://humanityplus.org/.

Hume, R.E., Trans. 1934. *The Thirteen Principal Upanishads*. London: Oxford University Press.

Hutchinson, A. 2018. *Mind, Body and the Curiously Elastic Limits of Human Performance*. New Zealand: HarperCollins Publishers.

Huxley, Julian. 1957. *New Bottles for New Wine*. London: Chatto & Windus.

Huxtable, Richard. 2018. Cryonics in the Courtroom: Which Interests? Whose Interests? *Medical Law Review* 26 (3): 476–499.

Ishikawa, Maria. 2018. Mindfulness in Western Contexts Perpetuates Oppressive Realities for Minority Cultures: The Consequences of Cultural Appropriation. *Simon Fraser University Educational Review* 11(1). Accessed May 28, 2020. https://doi.org/10.21810/sfuer.v11i1.757.

Jackson, Roy. 2020. *Muslim and Supermuslim: The Quest for the Perfect Being and Beyond.* In *Palgrave Studies in the Future of Humanity and Its Successors*, ed. Calvin Mercer and Steve Fuller. New York: Palgrave Macmillan.

Jackson, Susan, and Mihalyi Csikszentmihalyi. 1999. *Flow in Sports: The Keys to Optimal Experiences and Performance.* Champaign, IL: Human Kinetics.

Jesse, Robert. 1996. Entheogens: A Brief History of Their Spiritual Use. *Tricycle: The Buddhist Review* (Fall).

Jones, D. Gareth. 2013. Moral Enhancement as a Technological Imperative. *Perspectives on Science and Christian Faith* 65: 187–195.

Joy, Bill. 2000. Why the Future Doesn't Need Us. *Wired.* April.

Kahn, Peter. 2016. Bioethics, Religion, and Public Policy: Intersections, Interactions, and Solutions. *Journal of Religion and Health* 55 (5): 1546–1560.

Kaku, Michio. 2014. *The Future of the Mind: The Scientific Quest to Understand, Enhance, and Empower the Mind.* New York: Doubleday.

Kass, Leon. 2001. L'Chaim and Its Limits: Why Not Immortality? *First Things* (May). https://www.firstthings.com/article/2001/05/lchaim-and-its-limits-why-not-immortality.

Kim, Grace Ji-Sun, and Susan M. Shaw. 2017. Intersectional Theology: A Prophetic Call for Change. *Huffington Post*, March 31. http://www.huffingtonpost.com/entry/intersectional-theology-a-prophetic-call-forchange_us_58dd823de4b0fa4c09598794.

KiroRus. n.d.. https://kriorus.ru/en.

Kissinger, Henry. 2018. How the Enlightenment Ends: Philosophically, Intellectually—In Every Way—Human Society is Unprepared for the Rise of Artificial Intelligence. *The Atlantic.* June.

Knibbs, Kate. 2020. There's No Cure for Covid-19 Loneliness, but Robots Can Help. *Wired Magazine* (June 22).

Kobar, S.E., et al. 2017. Ability to Gain Control Over One's Own Brain Activity and its Relation to Spiritual Practice: A Multimodal Imaging Study in Frontiers. *Human Neuroscience.* https://doi.org/10.3389/fnhum.2017.00271.

Krüger, Oliver. 2015. The Suspension of Death. The Cryonic Utopia in the Context of the U.S. Funeral Culture. *Marburg Journal of Religion* 15 (1): 1–19.

Kurzweil, Ray. 1999. *The Age of Spiritual Machines: When Computers Exceed Human Intelligence.* New York: Penguin.

———. 2005. *The Singularity Is Near: When Humans Transcend Biology.* New York: Viking.

———. 2012. *How to Create a Mind.* New York: Viking.

Kurzweil, Ray, and Neil Degrasse. 2016. *2029 Singularity Year—Neil Degrasse Tyson & Ray Kurzweil.* April 21. https://www.youtube.com/watch?v=EyFYFjESkWU.

Kurzweil, Ray, and Terry Grossman. 2004. *Fantastic Voyage: The Science Behind Radical Life Extension, Live Long Enough to Live Forever.* Emmaus, PA.

"Kurzweil: Accelerating Intelligence." n.d.. www.KurzweilAI.net.

Kushner, Harold. 1981. *Why Bad Things Happen to Good People.* New York: Anchor Books.

Laderman, Gary. 1996. *The Sacred Remains. American Attitudes towards Death, 1799–1883*. New Haven: Yale University.
Landua, Iddo. 2018. The Case for Technological Mysticism. *Journal of Posthuman Studies* 2 (1): 67–85.
Latorra, Michael. 2015. What Is Buddhist Transhumanism? *Theology and Science* 13 (2): 219–229.
Lazzari, Edmund Michael. 2017. Would St. Thomas Aquinas Baptize an Extraterrestrial? *New Blackfriars*: 440–457. https://doi.org/10.1111/nbfr.12319.
Lee, Newton. 2019. *The Transhumanism Handbook*. New York: Springer.
Leidenhag, Joanna. 2016. A Critique of Emergent Theologies. *Zygon: A Journal of Religion & Science* 51 (4): 867–882. https://doi.org/10.1111/zygo.12300.
Levy, David. 2006. *Robots Unlimited: Life in a Virtual Age*. Wellesley, MA: A. K. Peters.
Life Extension Foundation. n.d.. www.lifeextension.com.
Linden, Ingemar Patrick. forthcoming. *The Case Against Death*. MIT Press.
Lindsey, Hal. 1970. *The Late, Great Planet Earth*. Grand Rapids, MI: Zondervon.
———. 1973. *There's a New World Coming*. Grand Rapids, MI: Zondervon.
———. 1982. *The 1980s: Countdown to Armageddon*. New York: Westgate.
"Living to 120 and Beyond: Americans' Views on Aging, Medical Advances and Radical Life Extension." 2013. Pew Research Center (Washington: Pew Research Center), 9, 13.
Locke, John. 1689. *Essay Concerning Human Understanding*. http://www.earlymoderntexts.com/assets/pdfs/locke1690book2.pdf.
Loene, Randal A. 2014. Feasible Mind Uploading. In *Intelligence Unbound: The Future of Uploaded and Machine Minds*, ed. Russell Blackford and Damien Broderick, 90–102. West Sussex, UK: Wiley Blackwell.
Logan, Robert K., and Adriana Braga. 2020. *AI and the Singularity: A Fallacy or Great Opportunity?* Basel: MDPI.
Lomax, J.W., J.J. Kripal, and K.I. Pargament. 2011. Perspectives on Sacred Moments in Psychotherapy. *The American Journal of Psychiatry* 168 (1): 12–18.
Lorrimar, Victoria. 2019. Mind Uploading and Embodied Cognition: A Theological Response. *Zygon* 54, no. 1 (3/19): 191–206.
MacCracken, Harry and Lev Grossman. 2013. Can Google Solve Death? *Time* (September 30). http://content.time.com/time/covers/0166412013093000.html.
MacFarlane, James Michael. 2020. *Transhumanism as a New Social Movement: The Techno-Centred Imagination*. In *Palgrave Studies in the Future of Humanity and Its Successors*, ed. Calvin Mercer and Steve Fuller. New York: Palgrave Macmillan.
Mackler, A. 2003. *Introduction to Jewish and Catholic Bioethics*. Washington, DC: Georgetown University Press.
Maher, Derek. 2009. Two Wings of a Bird. In *Religion and the Implications of Radical Life Extension*, Co-edited by Derek Maher and Calvin Mercer. New York: Palgrave Macmillan; Republished in Paperback, 2014.
Maher, Derek, and Calvin Mercer, eds. 2009. *Religion and the Implications of Radical Life Extension*. New York: Palgrave Macmillan. Republished Paperback, 2014.
Mahoney, A., K.I. Pargament, B. Cole, T. Jewell, and R. Phillips. 2005. A Higher Purpose: The Sanctification of Strivings in a Community Sample. *The International Journal for the Psychology of Religion* 15 (3): 239–262.
Majlesi, Mohammad-Baqer. n.d. *Bihar al-Anwar*, vol. 1, 177.

Marcuse, Herbert. 1964. *One Dimensional Man: Studies in the Ideology of Advanced Industrial Society*. Boston: Beacon Press.

McFee, Daniel. 2015. The Risks of Transhumanism: Religious Engagements with the Precautionary and Proactionary Principle. In *Religion and Transhumanism: The Unknown Future of Human Enhancement*, ed. Calvin Mercer and Tracy J. Trothen, 217–228. Westport, CT: Praeger.

Mercer, Calvin. 2008. Cryonics and Religion: Friends or Foes? *Cryonics* 29 (1): 10–21.

———. 2009. *Slaves to Faith: A Therapist Looks Inside the Fundamentalist Mind*. Westport, CT: Praeger.

———. 2014. Insisting on Soma in the Debate about Radical Life Extension: One Protestant's Perspective. In *Transhumanism and the Body: The World Religions Speak*, co-edited Calvin Mercer and Derek Maher. In *Palgrave Studies in the Future of Humanity and Its Successors*, Series Co-edited by Calvin Mercer and Steve Fuller. New York: Palgrave Macmillan.

———. 2015a. Bodies and Persons: Theological Reflections on Transhumanism. *Dialog* 54 (1): 27–33.

———. 2015b. Whole Brain Emulation Requires Enhanced Theology, and a 'Handmaiden'. *Theology and Science* 13 (2): 175–186.

———. 2017a. A Theological Assessment of Whole Brain Emulation: On the Path to Superintelligence. In *Religion and Human Enhancement: Death, Values, and Morality*, ed. Tracy Trothen and Calvin Mercer, 89–104. In *Palgrave Studies in the Future of Humanity and Its Successors*, Series Editors, Calvin Mercer and Steve Fuller. New York: Palgrave Macmillan.

———. 2017b. Resurrection of the Body and Cryonics. *Religions* 8/5, 96 (May): 1–9. www.mdpi.com/journal/religions, https://doi.org/10.3390/rel8050096. In Special Issue "Religion and the New Technologies." http://www.mdpi.com/journal/religions/special_issues/new_technologies.

———. 2020b. A Theological Embrace of Transhuman and Posthuman Beings. *Perspectives on Science and Christian Faith* 72 (2): 83–88.

Mercer, Calvin, and Thomas Durham. 1999. Religious Mysticism and Gender Orientation. *Journal for the Scientific Study of Religion* 38 (1): 175–182.

Mercer, Calvin, and Derek Maher, Eds. 2014. *Transhumanism and the Body: The World Religions Speak*. In *Palgrave Studies in the Future of Humanity and Its Successors*. Series Editors Calvin Mercer and Steve Fuller. New York: Palgrave Macmillan.

Miah, A. 2010. The DREAM Gene for the Posthuman Athlete: Reducing Exercise-induced Pain Sensations Using Gene Transfer. In *The Anthropology of Sport and Human Movement: A Biocultural Perspective*, ed. R.R. Sands and L.R. Sands, 327–341. Plymouth, UK: Lexington Books.

Mittal, Nitish, et al. 2017. The Gcn4 Transcription Factor Reduces Protein Synthesis Capacity and Extends Yeast Lifespan. *Nature Communications* 8 (1). https://doi.org/10.1038/s41467-017-00539-y.

Molloy, Michael. 2020. *Experiencing the World's Religions: Tradition, Challenge, and Change*. 8th ed. New York: McGraw-Hill.

Monod, Stefanie, Estelle Martin, Brenda Spencer, Etienne Rochat, and Christophe Büla. 2012. Validation of the Spiritual Distress Assessment Tool in Older Hospitalized Patients. *BMC Geriatrics* 12 (13): 1–9.

Moravec, Hans. 1988. *Mind Children: The Future of Robot and Human Intelligence*. Cambridge, MA: Harvard University.

More, Max. 2013. The Philosophy of Transhumanism. In *The Transhumanist Reader: Classical and Contemporary Essays on the Science, Technology, and Philosophy of the Human Future*, ed. Max More and Natasha Vita-More, 3–17. West Sussex: Wiley-Blackwell.

———. n.d. A Letter to Mother Nature: Amendments to the Human Constitution. http://strategicphilosophy.blogspot.com/2009/05/its-about-ten-years-since-i-wrote.html.

Moritz, Joshua M. 2011. Evolution, the End of Human Uniqueness, and the Election of the Imago Dei. *Theology and Science* 9 (3): 307–339.

Murphy, Mike. 2017. Stephen Hawking: AI Could be Best—or Worst—Thing in Human History. *Market Watch*, November 17. https://www.marketwatch.com/story/stephen-hawking-ai-could-be-best-or-worst-thing-in-human-history-2017-11-06.

Musa, Aisha Y. 2009. A Thousand Years, Less Fifty: Toward a Quranic View of Extreme Longevity. In *Religion and the Implications of Radical Life Extension*, Co-edited by Derek Maher and Calvin Mercer, 124–31. New York: Palgrave Macmillan; Republished in Paperback, 2014.

Musk, Elon. n.d. Elon Musk's Last Warning About Artificial Intelligence. https://www.youtube.com/watch?v=B-Osn1gMNtw&feature=youtu.be.

Muzyka, Kamil. 2013. The Outline of Personhood Law Regarding Artificial Intelligences and Emulated Human Entities. *Journal of Artificial General Intelligence* 4 (3): 164–169. https://doi.org/10.2478/jagi-2013-0010. ISSN 1946-0163.

Nagoshi, Craig T., and Julie L. Nagoshi. 2012. Being Human versus Being Transhuman: The Mind-Body Problem and Lived Experience. In *Building Better Humans? Refocusing the Debate on Transhumanism*, ed. Hava Tirosh-Samuelson and Kenneth L. Mossman, 303–319. Berlin: Peter Lang.

Ni, Zhange. 2020. Reimagining Daoist Alchemy, Decolonizing Transhumanism: The Fantasy of Immortality Cultivation in Twenty-First Century China. *Zygon: Journal of Religion and Science* 55 (3): 748–771. https://doi.org/10.1111/zygo.12634.

Nick, Bostrom, and Anders Sandberg. 2009. Cognitive Enhancement: Methods, Ethics, Regulatory Challenges. *Science and Engineering Ethics* 15: 311–341. https://doi.org/10.1007/s11948-009-9142-5.

Noss, David S., and Blake R. Grangaard. 2017. *A History of the World's Religions.* 14th ed. Abingdon: Taylor and Francis.

O'Callaghan, Sean. 2017. Technological Apocalypse: Transhumanism as an End-Time Religious Movement. In *Religion and Human Enhancement: Death, Values, and Morality*, ed. Tracy Trothen and Calvin Mercer, 67–88. In *Palgrave Studies in the Future of Humanity and Its Successors*, ed. Calvin Mercer and Steve Fuller. New York: Palgrave Macmillan.

Ohlheiser, Abby. 2014. Pope Francis Says He Would Definitely Baptize Aliens If They Asked Him To. *The Atlantic.* May 12. https://www.theatlantic.com/international/archive/2014/05/pope-francis-says-he-would-definitely-baptize-aliens-if-they-wanted-it/362106/.

"Old Age is Over! If You Want It." 2019. *MIT Technology Review* 122, no. 5 (September/October).

Olsen, Anders, Maithili C. Vantipalli, and Gordon J. Lithgow. 2006. Using Caenorhabditis Elegans as a Model for Aging and Age-Related Diseases. *Annals of the New York Academy of Sciences* 1067 (1): 120–128. https://doi.org/10.1196/annals.1354.015.

Olshansky, S. Jay. 2013. Reinventing Aging: An Update on the Longevity Dividend. *Aging Today: A Bimonthly Newspaper of the American Society on Aging* (March/April). https://www.asaging.org/blog/reinventing-aging-update-longevity-dividend.

Oregon Cryonics. n.d.. http://www.oregoncryo.com/index.html.

Ormond, K.E., et al. 2017. Human Germline Genome Editing. *American Journal of Human Genetics* 101 (2): 167–176.

Panch, Trishan, H. Mattie, and R. Atun. 2019. Artificial Intelligence and Algorithmic Bias: Implications for Health Systems. *Journal of Global Health* 9 (2). https://doi.org/10.7189/jogh.09.020318.

Parens, Erik, and Lori P. Knowles. 2003. *A Special Supplement to the Hastings Center Report: Reprogenetics and Public Policy: Reflections and Recommendations.* July–August. https://live-the-hastings-center.pantheon.io/wp-content/uploads/reprogenetics_and_public_policy.pdf.

Parfit, Derek. 1984. *Reasons and Persons.* Oxford: Clarendon Press.

Pargament, K.I. 2007. *Spiritually Integrated Psychotherapy: Understanding and Addressing the Sacred.* New York: Guilford.

Pargament, K.I., D. Oman, J. Pomerleau, and A. Mahoney. 2017. Some Contributions of a Psychological Approach to the Study of the Sacred. *Religion* 47 (4): 718–744.

Parks, J., and V. Wike. 2010. Theories and Values in Bioethics. In *Bioethics in a Changing World*, ed. J. Parks and V. Wike, 1–28. New Jersey: Prentice Hall, Pearson Education, Inc.

Parry, Jim. 2005. Must Scientists Think Philosophically About Science? In *Philosophy and the Sciences of Exercise, Health and Sport: Critical Perspectives on Research Methods*, ed. Mike McNamee, 20–31. New York: Routledge.

Pederson, Isabel, and Andrew Iliadis, eds. 2020. *Embodied Computing—Wearables, Implantables, Embeddables, Ingestibles.* Boston: MIT.

Persson, Ingmar, and Julian Savulescu. 2008. The Perils of Cognitive Enhancement and the Urgent Imperative to Enhance the Moral Character of Humanity. *Journal of Applied Philosophy* 25: 162–177.

———. 2012. *Unfit for the Future: The Need for Moral Enhancement.* Oxford, UK: Oxford University Press.

———. 2019. The Evolution of Moral Progress and Biomedical Moral Enhancement. *Bioethics* 33 (7): 814-819.

Peters, Ted. 1996. *Playing God: Genetic Determinism and Human Freedom.* New York: Routledge.

———. 2003a. *Science, Theology, and Ethics.* Burlington, VT: Ashgate.

———. 2003b. *Playing God? Genetic Determinism and Human Freedom.* 2nd ed. New York: Routledge.

———. 2006. *Anticipating Omega: Science, Faith, and Our Ultimate Future.* In *Religion, Theology and Natural Science*, vol. 7. Gottingen: Vandenhoeck & Ruprecht.

———. 2009. Reflections on Radical Life Extension. In *Religion and the Implications of Radical Life Extension*, ed. Derek Maher and Calvin Mercer. New York: Palgrave Macmillan; Republished in Paperback, 2014.

———. 2019. The Ebullient Transhumanist and the Sober Theologian. *Scientia et Fides*, July 2: 97–117.

Peters, Ted, Robert John Russell, and Michael Welker, eds. 2002. *Resurrection: Theological and Scientific Assessments.* Grand Rapids: William B. Eerdmans.

Pigliucci, Massimo. 2014. Mind Uploading: A Philosophical Counter-Analysis. In *Intelligence Unbound: The Future of Uploaded and Machine Minds*, ed. Russell Blackford and Damien Broderick, 119–131. West Sussex, UK: Wiley Blackwell.

Piore, A. 2015. A Shocking Way to Fix the Brain. *MIT Technology Review* (October 8). https://www.technologyreview.com/s/542176/a-shocking-way-to-fix-the-brain/.

Pitman, R.K., et al. 2002. Pilot Study of Secondary Prevention of Post-traumatic Stress Disorder with Propranolol. *Biological Psychiatry* 51 (2): 189–192.

Plato. 1921. *Cratylus*. In *Plato in Twelve Volumes*, vol. 12. Trans. Harold N. Fowler. Cambridge, MA: Harvard University. http://www.perseus.tufts.edu/hopper/text?doc=Perseus%3Atext%3A1999.01.0172%3Atext%3DCrat.%3Asection%3D402a.

Popper, Karl. 1978. Natural Selection and the Emergence of Mind. *Dialectica* 32: 339–355.

Porter, J. 2009. Implicit Religion in Popular Culture: The Religious Dimensions of Fan Communities. *Implicit Religion* 12 (3): 271–280.

Post, Stephen G. 2005. Altruism, Happiness, and Health: It's Good to Be Good. *International Journal of Behavioral Medicine* 12 (2): 66–77. https://doi.org/10.1207/s15327558ijbm1202_4.

Prisco, Giulio. 2012. Transhumanist Religion 2.0. *Kurzweil Accelerating Intelligence. Daily Blog* (July 13). https://www.kurzweilai.net/transhumanist-religion-2-0.

Purtill, C. 2019. The Robot Will Help You Now: How They Could Fill the Staffing Gaps in the Eldercare Industry. *TIME Magazine* (November 4).

Rachels, James. 2003. *Elements of Moral Philosophy*. 4th ed. Boston: McGraw-Hill.

Raveh, Anat Ringel, and Boaz Tamir. 2020. *AI and the Singularity: A Fallacy or Great Opportunity?* Edited by Robert K. Logan and Adriana Braga, 197–215. Basel: MDPI.

Reed, Randall. 2018. A New Pantheon: Artificial Intelligence and 'Her'. *Journal of Religion and Film* 22 (2). https://works.bepress.com/randall-reed/1/download/.

Reydon, Thomas A.C. 2019. Philosophy of Technology. *Internet Encyclopedia of Philosophy*. Accessed November 2019. https://www.iep.utm.edu/technolo/.

Reynolds, Thomas E. 2008. *Vulnerable Communion: A Theology of Disability and Hospitality Grand Rapids*. Michigan: Brazos Press.

Richards, Jay W. 2002. *Are We Spiritual Machines? Ray Kurzweil vs. the Critics of Strong AI*. Seattle: Discovery Institute.

Rigveda 8.48.3. 1896. *The Hymns of the Rig Veda*. Translated by Ralph T. H. Griffith. Benares, India: E. J. Lazarus and Company.

Ritchie, Barry G. 2012. The (Un)Likelihood of a High-Tech Path to Immortality. In *Building Better Humans? Refocusing the Debate on Transhumanism*, vol. 3, ed. Hava Tirosh-Samuelson and Kenneth L. Mossman. In *Beyond Humanism: Trans- and Posthumanism*, ed. Stefan Lorenz Sorgner. Frankfurt: Peter Lang.

Robinson, Peter. 2019. Fixed Points in a Changing World. In *Spiritualities, Ethics, and Implications of Human Enhancement and Artificial Intelligence*, ed. Chris Hrynkow, 227–241. Delaware: Vernon Press.

Rothblatt, Martine. 2014. *Virtually Human: The Promise—And Peril—Of Digital Immortality*. New York: St. Martin's.

Rudman, Stanley. 1997. *Concepts of Person and Christian Ethics*. Cambridge: Cambridge University.

Ryle, Gilbert. 1949. *The Concept of Mind*. Chicago: University of Chicago Press.

Sajoo, A.B. 2014. Negotiating Virtue: Principlism and *Maslaha* in Muslim Bioethics. *Studies in Religion* 43 (1): 53–69.

Sandberg, Anders. 2013. Morphological Freedom—Why We Not Just Want It, but Need It. In *The Transhumanist Reader: Classical and Contemporary Essays on the Science, Technology, and Philosophy of the Human Future*, ed. Max More and Natasha Vita-More, 56–64. West Sussex, UK: Wiley-Blackwell.

Saslow, Laura Rose, et al. 2013. The Social Significance of Spirituality: New Perspectives on the Compassion–Altruism Relationship. *Psychology of Religion and Spirituality* 5: 201–218.

Scheidt, Hannah. 2015. The Fleshless Future: A Phenomenological Perspective on Mind Uploading. In *Religion and Transhumanism: The Unknown Future of Human Enhancement*, ed. Calvin Mercer and Tracy J. Trothen, 315–328. Santa Barbara: Praeger.

Schultz, A.M., and P.E. Carron. 2013. Socratic Meditation and Emotional Self-Regulation: Human Dignity in a Technological Age. *Journal of Interdisciplinary Studies* 25 (1–2): 137–160.

"Scientists' Open Letter on Cryonics." n.d.. https://www.biostasis.com/scientists-open-letter-on-cryonics/.

Seger, E. 2017. FDA Approves First Gene Therapy for Leukemia Treatment: The How's, Why's, Promise, and Peril. *The Science Distillery* (September 5). https://sciencedistillery.com/2017/09/05/fda-approves-first-gene-therapy-for-leukemia-treatment-the-hows-whys-promise-and-peril/.

Setta, S.M., and S.D. Shemie. 2015. An Explanation and Analysis of How World Religions Formulate Their Ethical Decisions on Withdrawing Treatment and Determining Death. *Philosophy, Ethics, and Humanities in Medicine* 10 (6). https://doi.org/10.1186/s13010-015-0025-x.

Shao, Grace. 2019. Medical Cannabis is Gaining Momentum in Asia. *CNBC* (July 14). https://www.cnbc.com/2019/07/15/medical-cannabis-is-gaining-momentum-in-asia.html.

Sharma, Arvind. 2009. 'May You Live Long:' Religious Implications of Extreme Longevity in Hinduism. In *Religion and the Implications of Radical Life Extension*, Co-editors Calvin Mercer and Derek Maher, 145–52. New York: Palgrave Macmillan; Republished in Paperback, 2014.

Shoemaker, David. 2005. Personal Identity and Ethics. Section 2.3. http://plato.stanford.edu/archives/spr2014/entries/identity-ethics/.

"Should We Pursue Genetic Cognitive Enhancement?" 2018. *The Hastings Center News* (April 3). https://www.thehastingscenter.org/news/enhanced-human-risks-opportunities/.

Skidmore, C.R., E.A. Kaufman, and S.E. Crowell. 2016. Substance Use Among College Students. *Child and Adolescent Psychiatric Clinics in North America* 25: 735–753. https://doi.org/10.1016/j.chc.2016.06.004.

"Slaughterbots." n.d.. https://www.youtube.com/watch?v=9CO6M2HsoIA.

Steinhard, Eric. 2008. Teilhard de Chardin and Transhumanism. *Journal of Evolution and Technology* 20: 1–22.

Swinton, John. 2014. What the Body Remembers: Theological Reflections on Dementia. *Journal of Religion, Spirituality & Aging* 26: 160–172.

Sysiak, Pawel. 2016. When Will the First Machine Become Superintelligent? *Medium*. April 11.

Tanz, Jason. 2016. The End of Code: Soon We Won't Program Computers. We'll Train Them Like Dogs. *Wired* (May 17). https://www.wired.com/2016/05/the-end-of-code/.

Tennison, M.N., and Moreno, J.D. 2012. Neuroscience, Ethics, and National Security: The State of the Art. *PLOS* (March 20). http://journals.plos.org/plosbiology/article?id=10.1371/journal.pbio.1001289.

"The Brain Stimulator: Stimulate Your Life." n.d.. https://thebrainstimulator.net/what-is-tdcs/.

The President's Council on Bioethics, Beyond Therapy: Biotechnology and the Pursuit of Happiness. October 2003. https://bioethicsarchive.georgetown.edu/pcbe/reports/beyondtherapy/chapter4.html.

The Works of Benjamin Franklin. 1882. Vol. 6. Edited by Jared Sparks. Chicago: Townsend Mac Coun.

"The World's Most Dangerous Ideas." 2004. *Foreign Policy* 83 (5), September/October.

Thompson, Clive. 2016. Sure, A.I. Is Powerful—But Can We Make It Accountable? *Wired* (October 27). https://www.wired.com/2016/10/understanding-artificial-intelligence-decisions/.

Thweatt-Bates, J. Jeanine. 2011. Artificial Wombs and Cyborg Births. In *Transhumanism and Transcendence: Christian Hope in an Age of Technological Enhancement*, ed. Ron Cole-Turner, 101–114. Washington, DC: Georgetown University.

Thweatt-Bates, Jeanine. 2012. *Cyborg Selves: A Theological Anthropology of the Posthuman.* Burlington, VT: Ashgate Publishing.

———. 2015. Cindi, Six, and Her: Gender Relationality and Friendly Artificial Intelligence. In *Religion and Transhumanism: The Unknown Future of Human Enhancement*, ed. Calvin Mercer and Tracy J. Trothen, 39–50. Westport, CT: Praeger.

Tillich, Paul. 1953. *Systematic Theology.* Vol. 2. Chicago: University of Chicago.

Trancik, Emily K. 2013. Lost in Translation: Spiritual Assessment and the Religious Tradition. *Christian Bioethics* 19 (3): 282–298.

Trothen, T.J. 2015. *Winning the Race? Religion, Hope, and Reshaping the Sport Enhancement Debate*, Sport and Religion Series. Macon, GA: Mercer University Press.

———. 2017a. Christian Anthropology: Doing Good Through Science and Technology. In *Technology and the Image of God: A Canadian Conversation*, 18–19. Canadian Council of Churches, Faith and Life Sciences.

———. 2017b. Moral Bioenhancement Through an Intersectional Theo-Ethical Lens: Refocusing on Divine Image-Bearing and Interdependence. *Religions* 8 (5): 1–14. https://doi.org/10.3390/rel8050084.

———. 2017c. Moral Bioenhancement from the Margins: An Intersectional Christian Theological Reconsideration. In *Religion and Human Enhancement: Death, Values, and Morality*, ed. Tracy J. Trothen and Calvin Mercer, 245–266. Hampshire, UK: Palgrave Macmillan.

———. 2018. *Spirituality, Sport, and Doping: More than Just a Game.* SpringerBriefs Sport and Religion Series. Switzerland: Springer International Publishing.

Twigg, Robert C., Thomas Hengen, and Marlyn Bennett. 2008. Going Back to the Roots: Using the Medicine Wheel in the Healing Process. *First Peoples Child & Family Review* 4 (1): 10–19.

Universe Spirit. n.d. The Religion 2.0 Manifesto: Open Source Meta-Religion For the Twenty First Century. https://www.universespirit.org/what-is-universe-spirituality-the-universe-spirit-community-and-the-universe-spirituality-movement.

Vanlaere, Linus, Madeleine Timmerman, Marleen Stevens, and Chris Gastmans. 2012. An Explorative Study of Experiences of Healthcare Providers Posing as Simulated Care Receivers in a 'Care-Ethical' Lab. *Nursing Ethics* 19 (1): 68–79.

Vanlaere, Linus, Trees Coucke, and Chris Gastmans. 2010. Experiential Learning of Empathy in a Care-Ethics Lab. *Nursing Ethics* 17 (3): 325–336.

Vincent, James. 2017. Putin Says the Nation that Leads in AI 'Will be the Ruler of the World.' *The Verge*, September 4. https://www.theverge.com/2017/9/4/16251226/russia-ai-putin-rule-the-world.

Vita-More, Natasha. 2004. The New [Human] Genre—Primo Posthuman. Presentation at Ciber@RT Conference, Bilboa, Spain. http://www.natasha.cc/paper.htm.

———. 2008. "The Transhumanist Manifesto." https://natashavita-more.com/transhumanist-manifesto/.

———. 2011. The Posthuman Future—Interview with Natasha Vita-More. *Studio 360*, November 4. http://www.wnyc.org/story/233794-posthuman-future/.

———. 2012. *Life Expansion: Toward an Artistic Design-Based Theory of the Transhuman/Posthuman*. PhD diss. University of Plymouth. http://hdl.handle.net/10026.1/1182.

———. 2013a. Aesthetics: Bringing the Arts and Design into the Discussion of Transhumanism. In *The Transhumanist Reader: Classical and Contemporary Essays on the Science, Technology, and Philosophy of the Human Future*, ed. Max More and Natasha Vita-More, 18–27. West Sussex, UK: Wiley-Blackwell.

———. 2013b. Life Expansion Media. In *The Transhumanist Reader: Classical and Contemporary Essays on the Science, Technology, and Philosophy of the Human Future*, ed. Max More and Natasha Vita-More, 73–82. West Sussex, UK: Wiley-Blackwell.

———. 2014. Design of Life Expansion and the Human Mind. In *Intelligence Unbound: The Future of Uploaded and Machine Minds*, ed. Russell Blackford and Damien Broderick, 240–247. West Sussex, UK: Wiley Blackwell.

Wake Forest Institute for Regenerative Medicine. n.d.. https://school.wakehealth.edu/Research/Institutes-and-Centers/Wake-Forest-Institute-for-Regenerative-Medicine.

Waldrop, D.P. 2011. Denying and Defying Death: The Culture of Dying in 21st Century America. *The Gerontologist* 51 (4): 571–576.

Warwick, Kevin. 2014. The Cyborg Revolution. *NanoEthics* 8: 263–273. https://doi.org/10.1007/s11569-014-0212-z.

Waters, Brent. 2006. *From Human to Posthuman: Christian Theology and Technology in a Postmodern World*. In *Ashgate Science and Religion Series*, ed. Roger Trigg and J. Wentzel van Huyssteen. Burlington, VT: Ashgate Publishing.

Weintraub, David A. 2014. *Religions and Extraterrestrial Life: How Will We Deal With It?* New York: Springer.

Wellington, Naomi. 2014. Whole Brain Emulation: Invasive vs. Non-Invasive Methods. In *Intelligence Unbound: The Future of Uploaded and Machine Minds*, ed. Russell Blackford and Damien Broderick, 178–192. West Sussex, UK: Wiley Blackwell.

West, Traci C. 2019. *Solidarity and Defiant Spirituality: Africana Lessons on Religion, Racism, and Ending Gender Violence*. New York: New York University Press.

Wilkinson, David. 2013. *Science, Religion, and the Search for Extraterrestrial Intelligence*. Oxford: Oxford University.

Williams, Alex. 2019. Robots: Hot or Not? Love, Android Style, Sexy and Confusing. *The New York Times, SundayStyles* (January 1): 1, 8.

Wiseman, H. 2016. *The Myth of the Moral Brain: The Limits of Moral Enhancement*. Boston: MIT Press.

de Wolf, Aschwin, Brian Wowk, and Alcor Staff, eds. 2012. *Preserving Minds, Saving Lives: The Best Cryonics Writings of the Alcor Life Extension Foundation*. Scottsdale, AZ: Alcor Life Extension Foundation.

"Worshipping Immortality at the Church of Perpetual Life." April 11, 2016. https://www.youtube.com/watch?v=EvaC67CeBDA.

Yasinski, Emma. 2020. On the Road to 3-D Printed Organs. *The Scientist—Exploring Life, Inspiring Innovation* (February 26). https://www.the-scientist.com/news-opinion/on-the-road-to-3-d-printed-organs-67187.

Zhang, Chencheng, et al. 2020. An International Survey of Deep Brain Stimulation Utilization in Asia and Oceania: The DBS Think Tank East. *Frontiers in Human Neuroscience* 6 (July). https://www.frontiersin.org/articles/10.3389/fnhum.2020.00162/full.

Walsh, Brian. "Pope Francis Prays for Good AI." *Axios* (November 4, 2020). https://www.axios.com/pope-francis-good-ai-711e64fa-ef8a-4faa-afce-19c463f07425.html.

Index[1]

[1] Note: Page numbers followed by 'n' refer to notes.

C. Mercer, T. J. Trothen, *Religion and the Technological Future*, https://doi.org/10.1007/978-3-030-62359-3

The manufacturer's authorised representative in the EU is Springer
Nature Customer Service Centre GmbH, Europaplatz 3, 69115 Heidelberg,
Germany. If you have any concerns regarding our products, please
contact ProductSafety@springernature.com

Printed and bound by CPI Group (UK) Ltd, Croydon, CR0 4YY
30/09/2026
02225439-0017